Weaver's Practical Math for the Protective Coatings Industry

15835 Park Ten Place
Houston, TX 77084

800 Trumbell Drive
Pittsburgh, PA 15205

www.ampp.org

AMPP™

AMPP
Association for Materials Protection and Performance

Printed in the United States of America

Print ISBN 978-1-57590-475-7
eBook ISBN 978-1-57590-476-4

AMPP
Association for Materials Protection and Performance

15835 Park Ten Place
Houston, TX 77084

800 Trumbull Drive
Pittsburgh, PA 15205

ampp.org

Disclaimer

Raymond E.F. Weaver

Raymond E. F. Weaver was employed by SSPC beginning in 1972—first as the coordinator of SSPC research projects and coauthor of many technical reports and publications stemming from those projects and then segued into the area of standards development. He was also involved in preparing and grading the Protective Coating Specialist (PCS) examinations.

Mr. Weaver was a mathematics professor and chair of the mathematics department at the Community College of Allegheny County–Boyce Campus where he also taught physics from 1972 until 1984. He received his BS in Physics from the College of the Holy Cross and his MS in Physics from Carnegie-Mellon University.

Contents

PART I – REVIEW OF BASIC MATHEMATICS

Chapter 1 –Arithmetic **1**
1.1 Signed Numbers 1
1.2 Order of Operations 5
1.3 Fractions 7
1.4 Decimals 15
1.5 Exponents 16

Chapter 2 – Solving Simple Equations **21**
2.1 Linear Equations 21
2.2 Ratio and Proportion 25

Chapter 3 – Dimensional Analysis **33**
3.1 The Metric System 33
3.2 Conversion of Units 35
3.3 Deriving Paint Coverage Using Unit Conversions 40
3.4 Temperature Conversions 42

Chapter 4 – Geometry **47**
4.1 Area 47
4.2 Volume 55
4.3 Density 60

Chapter 5 – Computations **67**
5.1 Rounding 67
5.2 Significant Figures 67
5.3 Scientific Notation 72
5.4 Percents 75

Chapter 6 – Statistics **87**
6.1 Central Tendency 87
6.2 Dispersion 90

PART II – SPECIFIC COATINGS INDUSTRY APPLICATIONS

Chapter 7 – Thinning, Spreading Rate, and DFT **97**

7.1 Volume Solids – Relating Thinning, Spreading Rate, and DFT 97

7.2 Computing Theoretical Coverage 102

7.3 Effect of Thinning on Spreading Rate 106

7.4 Effect of Thinning on VOC Content 116

7.5 Computing DFT from Spreading Rate 119

7.6 Transfer Efficiency 121

7.7 Estimating Materials Needed 124

Chapter 8 – Concentration and Other Rates **131**

8.1 Pigment Volume Concentration (PVC) **131**

8.2 Computing Quantities From a Paint Composition Formula 132

8.3 Salt Concentrations 136

8.4 Volume Flow Rate 142

Chapter 9 – Making Judgments from Measurements and Computations **149**

9.1 Accuracy and Precision 149

9.2 DFT per SSPC-PA 2 150

9.3 Quality Assurance Acceptance Criteria 154

9.4 Dew Point and Relative Humidity 155

Chapter 10 – Cost Computations **161**

10.1 The Time Value of Money 161

10.2 Comparing the Cost of Coating Systems 168

Glossary **171**

Appendix A

List of Rules and Formulas 189

Appendix B
Conversion Factors—Metric/U.S. Customary 207

Appendix C
Temperature Conversion Chart—Celsius to Fahrenheit 213

Appendix D
Theoretical Spreading Rates 217

Appendix E
Dew Point and Relative Humidity Charts 219

Index 223

Introduction

Practical Math for the Protective Coatings Industry is specifically geared toward individuals who have avoided math since leaving school some time ago and now have a difficult time understanding the computations involved in the protective coatings field. It explains, for example, how to determine the wet film thickness (WFT) needed to meet a specified dry film thickness (DFT) when a paint is thinned by 10%. Those somewhat proficient in calculating quantities will find that this book offers insights into alternative approaches that will lead to a fuller understanding of mathematical relationships.

Sometimes the greatest obstacle to understanding mathematics lies in convincing yourself that you can do it. *Practical Math for the Protective Coatings Industry* uses real-world examples that you'll encounter on job sites every day to help you improve your math skills and thereby advance in your profession. Use Part 1 to review those principles underlying such applications as the effect of thinning on VOC content, pigment volume concentration, and dew point and relative humdity, which are discussed in Part 2. All sample computations come from SSPC standards and general industry practice.

Ray Weaver
SSPC Technical Advisor

PART I – REVIEW OF BASIC MATHEMATICS

Part I of this book reviews those basic mathematical principles underlying the formulas used every day in the protective coatings industry. This first part can also serve as a reference as formulas are developed and applications are presented in Part II.

Chapter 1 – Arithmetic

Objective: To review basic arithmetic, especially operations with signed numbers, decimals, fractions, and exponents.

1.1 Signed Numbers

Signed numbers are proceeded by plus (+) or minus (–) symbols that indicate whether or not they are negative or positive numbers. A number greater than zero is positive and can be written with a + sign, e.g., +8. A number less than zero is negative and is written with a – sign, e.g., –6. If a number is written without a sign, it is assumed to be positive. Zero is neither positive nor negative.

We encounter negative numbers often in our daily lives. When it is very cold, the temperature may be 4°F below zero. We express this number as –4°F. If the temperature increases by 4 degrees, we will be at zero. If I have a $20 balance in my checking account and I write a check for $50, I now have a negative balance of $30, or my new balance is –$30. I must deposit $30 to get my balance to zero. If I fail to make my bid (worth 300 points) in the first hand of a card game, my score is –300. I must score 300 more points to get back to zero.

For a complete discussion of signed numbers (i.e., positive and negative numbers), refer to an elementary algebra book or selected arithmetic text books. Here we present only a short summary of operations with signed numbers.

If we assign the positive direction to north, then south is the negative direction. Starting at zero (the origin) and moving 5 miles north, my new position is +5. If I start at the origin and go 6 miles south, my position will be negative 6, written as –6. My position, north or south of the origin, is indicated with a + or – sign. The opposite of +5 (5 miles north) is –5 (5 miles south). (+5) and (–5) are opposites of each other. Zero is its own opposite and is neither positive nor negative. A number and its **opposite** will always add up to zero.

Absolute Value: If I do not care about the direction, and only want to know how many miles I am from the origin, I want the absolute value or distance. Absolute value, also called the magnitude of a number, is the number written with a positive sign and is indicated with parallel lines on both sides of the number. The absolute value of positive 5 is written as $|+5| = 5$. The absolute value of negative 6 is written as $|-6| = 6$. Absolute value is always positive (or zero). The absolute value of zero is zero.

Rule for Adding Signed Numbers

- ☞ If the signs are the same, add the absolute values and keep the sign.
- ☞ If the signs are different, subtract the absolute values and keep the sign of the number with the larger absolute value.

Addition: Starting at zero (the origin) and moving 5 miles north, my new position is +5. If I then travel 3 more miles north, my position will be +8, i.e., I will be 8 miles north of the origin. If I then travel 10 miles south, I will be 2 miles south of the origin and my position will be –2. This can be expressed in a mathematical statement as:

$$(+5) + (+3) + (-10) = (-2)$$

The first step was to add the 5 and the 3 to get +8. The next step was to add the positive 8 and the negative 10 to get negative 2.

$$(+8) + (-10) = (-2)$$

If I continue 4 more miles south, I will be 6 miles south of the origin. This can be expressed as:

$$(-2) + (-4) = (-6)$$

In the example above with positive 8 added to negative 10, the difference between 8 and 10 is 2. Since 10 is larger than 8, we keep the sign of the 10, and the answer is negative 2.

Subtraction: Subtraction can be defined as adding the opposite. In each of these examples, the subtraction problem will be written as an addition problem with the same answer. Look them over and try to find the pattern.

Signs Written Explicitly

SUBTRACTION	ADDITION	WRITTEN "NORMALLY" WITHOUT EXTRA SIGNS
$(+7)-(+4) = (+3)$	$(+7)+(-4) = (+3)$	$7-4 = 3$
$(+9)-(+12) = (-3)$	$(+9)+(-12) = (-3)$	$9-12 = -3$
$(-5)-(+4) = (-9)$	$(-5)+(-4) = (-9)$	$-5-4 = -9$
$(+8)-(-2) = (+10)$	$(+8)+(+2) = (+10)$	$8-(-2) = 10$
$(-9)-(-5) = (-4)$	$(-9)+(+5) = (-4)$	$-9-(-5) = -4$

A simple rule to follow is "subtracting a negative is like adding a positive." We can think of the subtraction sign as canceling the negative sign.

The (–) sign really has three meanings:
1. An operator that tells us to subtract two numbers
2. Make the number negative
3. The opposite of a number

In the first example above $7 - 4 = 3$, we can think of the problem as subtracting 4 from 7; or we can think of it as adding a negative 4 to 7. In either case the result is 3.

We may find subtraction of negative numbers a bit difficult to "reason out" in our heads, so we simply learn the rules. Let us try this example to prove that the rules make sense, and then we can believe them. Assume I start with a balance of $60 in my checkbook and write my friend a check for $10 for the paint brush he gave me. My checkbook now has a balance of $50. Checks I write are debits and can be considered negative numbers. My friend refuses to accept the check and tears it up. I now have to subtract a debit (a negative number) from my balance. My adjusted checkbook balance is now $60. Subtracting a negative $10 is like adding a positive $10. This transaction can be written as:

$$50 - (-10) = 60$$

Rule for Subtracting Signed Numbers

☞ To subtract, add the opposite.

Multiplication and Division: Since multiplication is repetitive addition, we simply apply this definition to signed numbers. Three times negative four means:

$$3(-4) = (-4) + (-4) + (-4) = (-12)$$

according to the rules for addition. Continuing with this type of reasoning and using the fact that division is the inverse of multiplication, we arrive at the rule for multiplying and

Rule for Multiplying and Dividing Signed Numbers

☞ When multiplying or dividing two numbers with the same sign, the answer is positive.

☞ When multiplying or dividing two numbers with opposite signs, the answer is negative.

dividing signed numbers. A few examples follow.

$(+7)(+4) = (+28)$	$(+8) \div (+2) = (+4)$
$(+6)(-5) = (-30)$	$(+28) \div (-4) = (-7)$
$(-0.3)(+8) = (-2.4)$	$(-6) \div (+12) = (-0.5)$
$(-9)(-2) = (+18)$	$(-42) \div (-7) = (+6)$

If several minus signs precede a number, we work outward from the number.

$$-\left\{-[-(-8)]\right\} = -\left\{-[+8]\right\}$$

since $-(-8) = +8$. The opposite of negative 8 is positive 8. Continuing outward:

$$-\left\{-[+8]\right\} = -\{-8\}$$

The opposite of positive 8 is negative 8. The last step gives positive 8 because the opposite of negative 8 is positive 8.

$$-\{-8\} = +8$$

Notice that there were an even number (4) of negative signs. Pairs of negative signs cancel.

Multiplication can be indicated in several ways:

- With an x $\quad 5 \text{ x } 6 = 30$
- With a raised dot $\quad 5 \cdot 6 = 30$
- With parentheses $\quad (5)(6) = 30$

If there is no operation symbol (+, –, ÷, x or ·) between the factors, multiplication is assumed. For example, 6C means $6 \cdot C$; ab means $a \cdot b$.

1.2 Order of Operations

Jot down the answer to the problem:

$$5 + 6 \text{ x } 3 = ?$$

The two possibilities are 23 (multiply first) and 33 (add first), but only one is the correct answer. Mathematicians have agreed on the order in which the mathematical operations should be performed. Powers and roots are discussed briefly in Section 1.5.

In the original problem 5 + 6 x 3, there are no grouping symbols and no powers or roots. According to the rule, we should do the multiplication first, leaving us with 5 + 18. Lastly, we do the addition to get the correct answer of 23.

If for some reason we wanted to perform the addition first, then we would have to use grouping symbols. If we write (5 + 6) x 3, we do what is inside the parentheses first to get 11 x 3. Then we do the multiplication to get 33.

Order of Operations Rule

1. First perform operations within grouping symbols such as parentheses.
2. Evaluate powers and roots.
3. Do all multiplication and division in order from left to right. (A division may come before a multiplication.)
4. Do addition and subtraction in order from left to right. (A subtraction may come before an addition.)

Example: Evaluate 24 ÷ 6 x 2
Solution: Multiplying and dividing should occur in order from left to right, so we do the division first, getting 4 x 2 = 8.

Example: Evaluate this:
$7(2^3 - 3) - 4 \times 6$
Solution: The first step is to do what is inside the parentheses. We apply the order of operations rule to $2^3 - 3$ and do the exponent first, getting 8 – 3 = 5. The problem reduces to 7(5) – 4 x 6.
Next we do the multiplication to get 35 – 24.

Finally, we do the subtraction to get 11.

1.3 Fractions

A bridge consists of four equal spans. If three of these spans have deteriorated paint, then we can say that 3/4 of the bridge has deteriorated paint. We divided the bridge into four parts, and then chose three of them. The fraction 3/4 is made up of a numerator and a denominator. The number on top of the fraction is the numerator (in this case 3), and the number on the bottom is the denominator. The denominator can never be zero.

$$\frac{3}{4} \Leftrightarrow \frac{\text{numerator}}{\text{denominator}}$$

A Fraction as a Multiplication

$$\frac{a}{b} = a \cdot \frac{1}{b} \text{ OR } \frac{a}{b} = \frac{1}{b} \cdot a$$

If all four spans are deteriorated, then 4/4 of the bridge (or one whole bridge) would be bad. A basic principle of fractions is: **If the numerator and the denominator are equal, then the fraction is equal to one.** We will revisit this principle many times throughout this book.

Suppose none of the spans is in bad condition, then the fraction of the bridge that is bad is 0/4. Another basic principle of fractions is: **If the numerator of a fraction is zero, the whole fraction is equal to zero, regardless of the denominator.** The value of the fraction 0/0 is undetermined. Even if the numerator is zero, the denominator cannot be zero.

$$\frac{0}{\text{any non-zero number}} = 0$$

A Fraction as a Multiplication: A fraction can be considered a multiplication. If we multiply a single span by three, we get 3 spans. Since each span is 1/4 of the bridge, three spans would be 3 x 1/4, or 3/4 of the bridge. We can use letters "a" and "b" to represent the terms of a fraction.

A Fraction as a Division: A fraction can also be considered as a division. A different example would be more useful to illustrate this property of fractions. First do an easy division using whole numbers. Suppose we have 4 children and 20 candy kisses. How many would each child get if we divided the candy equally among them? This is quite easy to do as $20 \div 4 = 5$. So each child gets 5 candy kisses.

How do you divide 3 chocolate bars among 4 children? The schematic below shows that each child will get 3/4 of a bar. Thus, 3 divided by 4 is 3/4.

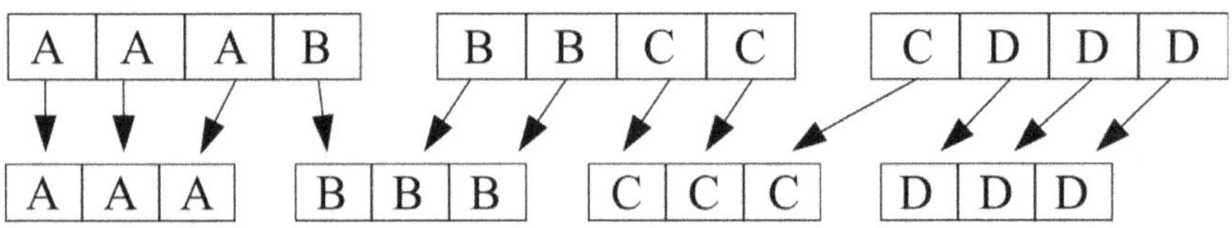

Remember to always divide BY the denominator. The denominator is the divisor.

Extending the idea of a fraction as a division, suppose the denominator is 1. Dividing by 1 does not change the value of a number. The next basic principle of fractions is: **Any number can be written as a fraction with a denominator of 1.** For example:

$$7 = 7/1,\ 12.8 = 12.8/1,\ 0.3 = 0.3/1, \text{etc.}$$

Converting Fractions: An **improper** fraction is a fraction where the numerator is not smaller than the denominator. Examples of improper fractions are 6/5, 4/3, 7/7, and 21/4.

A **mixed number** contains a whole number part and a fraction part: for example 2 1/5 or 6 3/8. When you have 3 1/2 gallons of paint, you have 3 whole gallons plus 1/2 gallon. The "+" sign is omitted when writing a mixed number.

To change an improper fraction to a mixed number simply divide the numerator by the denominator, and the remainder is the fractional part of the divisor.

For example:

$$\frac{11}{4} = 4\overline{)11}\ \text{Remainder } 3 \quad \text{OR} \quad \frac{11}{4} = 2\frac{3}{4}$$

$$\begin{array}{r} 2 \\ 4\overline{)11} \\ \underline{8} \\ 3 \end{array}$$

To change a mixed number to an improper fraction, multiply the whole number part by the denominator and add the numerator. This is the numerator of the improper fraction. The denominator is the same as in the mixed number. For example:

A Fraction as a Division

$$\frac{a}{b} = a \div b \quad \text{OR} \quad \frac{a}{b} = b\overline{)a}$$

$$2\frac{3}{8} = \frac{(2\times 8)+3}{8} = \frac{16+3}{8} = \frac{19}{8}$$

Changing the Terms of a Fraction: The "terms" of a fraction are simply the numbers used to write it. We know that if a pizza is cut into 8 pieces, 4 slices represent 4/8 of the pizza, which is the same as 1/2. Different terms are used to represent half the pizza. The rule that governs converting terms of a fraction is that any number can be multiplied by 1 without changing its value. The fraction c/c equals 1. If we multiply the fraction a/b by 1 we get the same fraction expressed in different terms.

$$\frac{a}{b}\cdot 1 = \frac{a}{b}\cdot\frac{c}{c} = \frac{a\cdot c}{b\cdot c}$$

In other words, we can multiply both the numerator and the denominator of a fraction by the same non-zero number without changing the value of the fraction.

To change 3/5 to 20[ths], we

multiply the numerator and the denominator by 4.

$$\frac{3}{5} = \frac{3 \times 4}{5 \times 4} = \frac{12}{20}$$

Reducing a Fraction to Lowest Terms: The process of raising a fraction to higher terms can be reversed by dividing the numerator and the denominator by the same number.

$$\frac{a}{b} = \frac{a \div c}{b \div c}$$

For example, 24/36 can be reduced to lower terms by dividing the numerator and the denominator by 2 to get 12/18. But 12 and 18 can both be divided by 2, resulting in 6/9. Continuing, 6 and 9 can both be divided by 3, giving 2/3. There is no number that will divide evenly into both 2 and 3, so 2/3 is considered to be reduced to lowest terms. Although 24/36 and 12/18 and 6/9 and 2/3 all represent the same number, the preferred way of expressing the number is 2/3, with the fraction in lowest terms.

We can make the denominator of a fraction anything we want if we multiply or divide the numerator and the denominator by the correct factor. The numerator, however, may turn out to be a decimal that we must round off.

Example: Change 3/4 to a fraction with a denominator of 1 and then to a fraction with a denominator of 6.

Solution: Divide both the numerator and the denominator by 4 as shown:

$$\frac{3}{4} = \frac{3 \div 4}{4 \div 4} = \frac{0.75}{1}$$

To convert this to a fraction with a denominator of 6, we then multiply the numerator and the denominator by 6.

$$\frac{0.75}{1} = \frac{0.75 \times 6}{1 \times 6} = \frac{4.5}{6}$$

Multiplying fractions: When we multiply fractions, we multiply the numerators and the denominators according to the formula:

$$\frac{a}{b} \cdot \frac{c}{d} = \frac{a \cdot c}{b \cdot d}$$

For example:

$$\frac{3}{5} \times \frac{7}{8} = \frac{3 \times 7}{5 \times 8} = \frac{21}{40}$$

If we multiply a number by its multiplicative inverse, the product is one. **Reciprocal** is another word for multiplicative inverse. The two numbers are reciprocals of each other. For example, 4/3 is the reciprocal of 3/4 because:

$$\frac{3}{4} \cdot \frac{4}{3} = \frac{12}{12} = 1$$

Similarly, 6 and 1/6 are reciprocals of each other because $6 \cdot 1/6 = 1$. We can find the reciprocal by interchanging the numerator and the denominator of a fraction.

Dividing fractions: To divide fractions you multiply by the reciprocal of the denominator. The reasoning behind this formula can be found in most arithmetic books.

$$\frac{a}{b} \div \frac{c}{d} = \frac{a}{b} \times \frac{d}{c}$$

The rule for division is often stated, "To divide fractions, invert the divisor and multiply." For example:

$$\frac{3}{8} \div \frac{5}{7} = \frac{3}{8} \times \frac{7}{5} = \frac{21}{40}$$

Adding Fractions: To add or subtract fractions with the same denominator, we add or subtract the numerators. For example:

$$\frac{3}{7} + \frac{2}{7} = \frac{5}{7} \quad \text{AND} \quad \frac{4}{5} - \frac{1}{5} = \frac{3}{5}$$

When adding and subtracting fractions, the denominators stay the same.

Adding and Subtracting Fractions with Different Denominators: Before adding or subtracting fractions with unlike denominators, it is necessary to convert the fractions so that they all have the same denominator. The least common denominator is the smallest number into which all the denominators will divide evenly. For example, to add 1/4 and 2/3, we have to find the smallest number that 4 and 3 will divide into evenly, namely 12. Then we must convert 1/4 and 2/3 into 12ths.

Multiplying 1/4 by 3/3, we can change 1/4 into 3/12. Similarly, multiplying 2/3 by 4/4 changes 2/3 into 8/12. The addition problem becomes:

$$\frac{1}{4}+\frac{2}{3}=$$

$$\frac{3}{12}+\frac{8}{12}=\frac{11}{12}$$

Example: Subtract 1/6 from 3/4.
Solution: The least common denominator is 12, so convert each fraction into 12^{ths}.

$$\frac{3}{4}-\frac{1}{6}=$$

$$\frac{9}{12}-\frac{2}{12}=\frac{7}{12}$$

Canceling: When multiplying fractions, common factors can be divided out (canceled) after performing the multiplication or prior to performing the multiplication. For example:

$$\frac{9}{16}\times\frac{2}{3}\times\frac{4}{5} = \begin{array}{c}72\\ \dfrac{\cancel{3}}{\cancel{10}}\\ 240\end{array}$$

which can be reduced to lowest terms by dividing numerator and denominator by 8 and then by 3 to get 3/10 as shown.

$$\frac{72\div 8}{240\div 8}=\frac{9}{30}=\frac{9\div 3}{30\div 3}=\frac{3}{10}$$

You could have reduced the fraction in one step by dividing by 24, but that is not as obvious a factor.

An easier method to perform this computation is to cancel common factors before multiplying. To cancel, divide any numerator and any denominator by the same number.

$$\frac{\overset{3}{\cancel{9}}}{\underset{\underset{2}{\cancel{8}}}{\cancel{16}}} \times \frac{\overset{1}{\cancel{2}}}{\underset{1}{\cancel{3}}} \times \frac{\overset{1}{\cancel{4}}}{5} = \frac{3}{10}$$

Using a Calculator to Compute with Fractions: Computations with fractions can be done on any calculator with an "a b/c" button. To enter a mixed number such as 5 7/8, enter the 5, push the a b/c button, enter the 7, push the a b/c button, and enter the 8. The display should read 5_|7_|8.

Example: Perform the operation using a calculator.

$$\frac{1}{2} \times \frac{3}{4} + 7\frac{3}{16}$$

Calculator solution: Enter the numbers in the following sequence.

ENTER	DISPLAY	COMMENT
1	1	
a b/c	1_|	to form the fraction
2	1_|2	
X	1_|2	to multiply
3	3	
a b/c	3_|	to form the fraction
4	3_|4	
+	3_|8	to add; subtotal is 3/8
7	7	
a b/c	7_|	to form the mixed number
3	7_|3	
a b/c	7_|3_|	to form the mixed number
16	7_|3_|16	
=	7_|9_|16	to get the final result: $7\frac{9}{16}$

1/2 x 3/4 gives 3/8. When 3/8 is added to 7 3/16, the calculator converts 3/8 to 6/16 and adds this to 7 3/16 to get the final answer of 7 9/16. The calculator will display the answer in lowest terms. By pushing the a b/c button repeatedly, some calculators will convert the answer to an improper fraction 121/16 or a decimal 7.5625.

A fraction bar is a grouping symbol, just like parentheses. Many formulas can be expressed as fractions with the numerator and/or the denominator consisting of one or more operations. After substituting into the formula, the resulting expression may look like:

$$\frac{(7)\cdot(5)}{(4)\cdot(9)}$$

which can be expressed as [(7) · (5)] ÷ [(4) · (9)]. After doing what is inside the square brackets first, we get 35 ÷ 36 or 35/36. The decimal equivalent is 0.97222…

Mistakes are often made when doing a problem like this on a calculator. We can use the parentheses on our calculator, but there is a simpler method involving fewer key strokes. Providing all operations are multiplication and division, the calculator steps are:

ENTER	DISPLAY	COMMENT
7	7	
X	7	to multiply
5	5	
÷	35	to divide; subtotal 7 x 5
4	4	this is DIVIDE not multiply; subtotal 35 ÷ 4
÷	8.75	
9	9	
=	0.9722222	final answer

To see why this sequence is correct, we can expand the problem as shown:

$$\frac{(7)\cdot(5)}{(4)\cdot(9)} = 7\cdot 5\cdot\frac{1}{4}\cdot\frac{1}{9} = 7\cdot 5\div 4\div 9$$

Multiplying by 1/4 is equivalent to dividing by 4, and multiplying by 1/9 is equivalent to dividing by 9.

1.4 Decimals

A decimal fraction is simply a fraction with a denominator of 10, 100, 1000, or some other multiple of 10. For example, 0.6 means 6/10 and 0.07 means 7/100 and .008 means 8/1000. The number 0.678 means 6/10 + 7/100 + 8/1000. If each of these fractions is raised to higher terms with a denominator of 1000, this becomes 600/1000 + 70/1000 + 8/1000, or 678/1000. Computations with decimals are easily done on a calculator, so discussion is omitted here.

When converting units within the metric system, we multiply or divide by powers of 10; and it is easier not to use a calculator because all we need to do is move the decimal point. How many places to move it and in what direction will be discussed in the section on conversion of units. The rules are simple when multiplying or dividing by multiples of 10.

To multiply a number by 10, move the decimal point one place to the **right**, i.e., make the number 10 times larger.

45.32 x 10 = 453.2
0.0067 x 10 = 0.067
456 x 10 = 4560

The decimal point in the third example is not written because it comes after the 6 in 456. To move it one place to the right in the last example requires adding a zero at the end of the number. Again, we do not write the decimal after the 0.

To multiply by 100, we do the same procedure as multiplying by 10 except that we move the decimal point two places to the right.

45.32 x 100 = 4532
0.0067 x 100 = 0.67
456 x 100 = 45,600

To multiply by 1000, move the decimal point three places to the right. To multiply by a million, move it six places, etc.

To divide by 10, move the decimal point one place to the **left**, making the number smaller.

$$45.32 \div 10 = 4.532$$
$$0.0067 \div 10 = 0.00067$$
$$456 \div 10 = 45.6$$

To divide by 100 move the decimal point two places to the left.

$$45.32 \div 100 = 0.4532$$
$$0.0067 \div 100 = 0.000067$$
$$456 \div 100 = 4.56$$

To divide by 1000 move the decimal point three places to the left. To divide by one million, move it six places to the left, etc.

1.5 Exponents

The rules for exponents are used in computing areas and volumes and in the conversion of units, e.g., converting cubic feet per minute (CFM) to cubic meters per minute. The short-hand way of expressing repetitive multiplication is to use exponents. The exponent (or power) is written as a superscript to the right of the number or letter to be multiplied by itself multiple times. For example:

$$4^3 = 4 \times 4 \times 4 \qquad \text{OR} \qquad m^3 = m \cdot m \cdot m$$

In these examples, the 4 and the m are the base, and 3 is the exponent or power.

An exponent can also be represented by a letter. For example,

$$6^n = [6 \times 6 \times 6 \times 6 \ldots \times 6]$$
n times

Example: The area of a square is calculated by squaring the length of a side, $A = s^2$. If the side is 6 ft, the area would then be $A = (6\text{ ft})^2 = (6\text{ ft}) \cdot (6\text{ ft}) = 36\text{ ft}^2$. In this type of problem, you must square both the 6 and the ft; the units as well as the numbers must be squared.

Powers of Negative Numbers: When a negative number is raised to a power, the sign of the answer depends on whether the exponent is even or odd. Let's consider powers of negative 1.

$(-1)^0 = +1$ any number to the zero power is positive 1
$(-1)^1 = -1$ an exponent of 1 need not be written
$(-1)^2 = +1$ $(-1)(-1)$ a negative times a negative is positive
$(-1)^3 = -1$ $(-1)(-1)(-1)$
$(-1)^4 = +1$ $(-1)(-1)(-1)(-1)$
$(-1)^5 = -1$ $(-1)(-1)(-1)(-1)(-1)$
and so on.

Negative one to an odd power is negative one, and negative one to an even power is positive one. This can be generalized to: a negative number to an odd power is a negative number, and a negative number to an even power is a positive number.

Examples: $(-2)^3 = (-8)$ and $(-2)^4 = (+16)$

Negative Exponents: A negative exponent means to take the reciprocal and then raise to the appropriate power.

$$4^{-1} = \frac{1}{4}$$

$$\left(\frac{1}{2}\right)^{-1} = \frac{2}{1} = 2$$

$$\left(\frac{5}{8}\right)^{-1} = \frac{8}{5}$$

$$\left(\frac{3}{4}\right)^{-2} = \left(\frac{4}{3}\right)^{2} = \frac{16}{9}$$

Notice that the number does not become negative because the exponent is negative.

Negative exponents are used when computing pH from the hydrogen ion concentration. They are also used extensively in the metric system to convert from one metric unit to another one.

Some negative powers of 10 are given below:

$10^{-1} = \frac{1}{10^1} = \frac{1}{10} = 0.1$
$10^{-2} = \frac{1}{10^2} = \frac{1}{100} = 0.01$
$10^{-3} = \frac{1}{10^3} = \frac{1}{1000} = 0.001$
$10^{-4} = \frac{1}{10^4} = \frac{1}{10{,}000} = 0.0001$

A negative exponent can be considered repetitive division. Multiplying by 10^{-2} is equivalent to dividing by 10 twice. For example:

$$463 \cdot 10^{-2} = 463 \cdot \frac{1}{10^2} = 463 \cdot \frac{1}{10 \cdot 10} = \frac{463}{10 \cdot 10} = \frac{463}{100} = 4.63$$

Roots: The inverse operation of raising a number to a power is finding a root. The square root of 25 is 5 because $5^2 = 25$. Using mathematical symbols this can be written as:

$$\sqrt{25} = 5$$

Similarly, the cube root of 8 is 2 because:

$$2^3 = 8 \text{ or}$$

$$\sqrt[3]{8} = 2$$

The nth root of a number, M, is designated as:

$$\sqrt[n]{M}$$

Consult an algebra book for a more detailed discussion of roots.

Some Rules for Exponents

- When the exponent is not written, it is assumed to be 1.

  $6^1 = 6$

  $b^1 = b$

- If the exponent is zero, the quantity equals 1.

  $6^0 = 1$

  $b^0 = 1$ for any b except 0. (0^0 is not defined.)

- When multiplying numbers with the same base, add the exponents.

  $2^3 \cdot 2^4 = 2^7$

  $b^m \cdot b^n = b^{m+n}$

- When dividing numbers with the same base, subtract the exponents.

  $2^5 \div 2^3 = 2^2$

  $b^m \div b^n = b^{m-n}$

- When taking a power to a power, multiply the exponents.

  $(2^3)^2 = 2^{3 \cdot 2} = 2^6 = 64$

  $(b^m)^n = b^{mn}$

- If the exponent is negative, take the reciprocal then raise to the appropriate power.

  $b^{-n} = \dfrac{1}{b^n}$

  $\left(\dfrac{a}{b}\right)^{-n} = \left(\dfrac{b}{a}\right)^n$

- Zero to any positive power is zero. Zero to a negative power and zero to the zero power are not defined.

  $0^n = 0 \ (n > 0)$

Chapter 2 – Solving Simple Equations

Objective: To review the procedures used to solve simple linear equations, including those associated with proportions.

2.1 Linear Equations

An equation is a statement of equality between two quantities or expressions. Usually, there is an unknown quantity represented by a letter called a variable. The object of solving an equation is to find the value for the variable that will make the statement true. For example, the equation $4n = 12$ is true only when $n = 3$. More complicated equations require a sequence of steps to get them into the form

$$\text{variable} = \text{number}$$

An equation is classified as "linear" if the exponent on the variable is 1.

If we consider an equation as a balance, we can maintain the balance as long as we do the same thing to both sides. We can add, subtract, multiply, or divide, as long as we add the same number to both sides, or subtract the same number from both sides, etc. The idea is to choose the operation that will simplify the equation. If multiplying both sides by the same number makes the equation more complicated, the resulting equation is not necessarily untrue; we should simply try a different operation. It is not always intuitive for everyone to know what operation to use. The following examples should expose a pattern.

Example: Solve $m - 8 = 12$
Solution: We want to get m by itself, without the 8. Since the 8 is subtracted from the m, we do the opposite of subtract and add 8 to both sides of the equation

$$m - 8 + \mathbf{8} = 12 + \mathbf{8}$$
$$m = 20$$

The solution (or root) to the equation is 20. That means that if m is replaced with 20, the equation is true. To check our solution, we use the original equation: (20) – 8 = 12, true.

Example: Solve N + 5 = 16
Solution: In order to get the letter, N, by itself, we must eliminate the 5. Since the left side of the equation has 5 added to N, we do the opposite of adding and subtract 5 from both sides.

$$N + 5 - \mathbf{5} = 16 - \mathbf{5}$$
$$N = 11$$

Example: Solve R/4 = 20
Solution: Since R/4 means R divided by 4, we do the opposite of divide and multiply both sides by 4.

$$\frac{R}{4} \cdot 4 = 20 \cdot 4$$
$$R = 80$$

Example: 9T = 27
Solution: Since T is multiplied by 9, we do the opposite and divide both sides by 9.

$$\frac{9T}{9} = \frac{27}{9}$$
$$T = 3$$

These examples were easy to solve by inspection without going through a seemingly needless procedure. However, we must understand the procedure in order to solve an equation such as 5.71S = 36.2. This answer is not at all obvious, and we should do it on a calculator. As in the example above, the variable is multiplied by a number, 5.71. Again, we do the opposite of multiplication and divide both sides by 5.71.

$$\frac{5.71S}{5.71} = \frac{36.2}{5.71}$$
$$S = 6.34$$

Our calculator gives 6.3397548 in the display, but we can round off as shown. How many digits to keep in our answer is discussed in Section 5.2 about significant figures.

Some equations require multiple steps. We can do several things in order to minimize the number of "careless" mistakes. First, it is good practice to keep the equal signs lined up. Second, many mistakes can be eliminated if we write down each step of the process, rather than performing multiple operations mentally. If we want to keep track of the particular operation we used to get from one step to the next, we can place notations at the end of each line as we work through to the solution. For example, if we add 6 to both sides, we can write A_6. Similarly, dividing by 4 is D_4, M_{-2} means multiply by negative 2, and S_8 is subtract 8.

Example: Solve $4n - 8 = n + 17$.
The strategy is to use addition and subtraction to get all the terms with the variable on one side of the equation and the numbers on the other side. We begin by subtracting n on both sides.

$$4n - \mathbf{n} - 8 = n - \mathbf{n} + 17$$

$$3n - 8 = 17$$ All terms with n are now on left.

$$3n - 8 + \mathbf{8} = 17 + \mathbf{8}$$ Add 8 to both sides.

$$3n = 25$$

Then divide both sides by 3.

$$\frac{3n}{\mathbf{3}} = \frac{25}{\mathbf{3}}$$

$$n = \frac{25}{3} = 8\frac{1}{3} \approx 8.33$$

As we become more proficient in solving equations, steps can be combined or completed mentally.

The **distributive property** of multiplication with respect to addition and subtraction is very useful in solving many equations and in understanding general algebraic manipulations. This property is a transformation between a multiplication problem and an addition problem. Using all letters, the distributive property can be stated:

$$a \cdot (b + c - d) = a \cdot b + a \cdot c - a \cdot d$$

If the order of operation rules (Section 1.2) are applied to the left hand side of this equation, we should do the adding and subtracting within the parentheses first and then multiply by a. This rule allows us to do the multiplication first (multiply each term inside the parentheses by a) and then add and subtract the products. A numerical example may better illustrate the distributive property. Simplifying both sides of the equation in accordance with the order of operation rules we get:

$$3\cdot(4+6-2) = 3\cdot4 + 3\cdot6 - 3\cdot2$$
$$3\cdot(10-2) = 12 + 18 - 6$$
$$3\cdot(8) = 30 - 6$$
$$24 = 24$$

Using both letters and numerals we would have:

$$5\cdot(x + 3y - 7) = 5x + 15y - 35$$

Steps for Solving Linear Equations

1. Remove parentheses using the distributive properties.
2. Combine like terms on each side.
3. Add and/or subtract to get all terms with variables on one side of the equation and all numbers on the other side.
4. Divide by the coefficient of the variable term.

Example: Solve this:
$7(x + 3) + 23 = 5(x - 2) - 4(5x - 6)$
Solution: We follow the steps given above.

$7x + 21 + 23 = 5x - 10 - 20x + 24$	Remove parentheses, Distributive property.
$7x + 44 = -15x + 14$	Combine like terms.
$7x + \mathbf{15x} + 44 - \mathbf{44} = -15x + \mathbf{15x} + 14 - \mathbf{44}$	Add 15x to both sides Subtract 44 from both sides.
$22x = -30$	
$\frac{22x}{22} = \frac{-30}{22}$	Divide by 22.
$x = -\frac{30}{22}$	
$x = -\frac{15}{11}$ or $-1\frac{4}{11}$	Reduce to lowest terms.

2.2 Ratio and Proportion

A ratio is the comparison of two numbers by division. The weight ratio of pigment to vehicle in a paint may be 2 to 1, which means that there is twice as much pigment as vehicle. The order in which a ratio is stated is important. For a gallon of paint with 12 lb of pigment and 6 lb of vehicle, the weight ratio of pigment to vehicle is:

$$\frac{12\,\cancel{\text{lb}}}{6\,\cancel{\text{lb}}} = \frac{2}{1} \text{ or } 2$$

Although it is not necessary to write the 1 in the denominator of a fraction, it is often written when describing a ratio. Notice in this example that the units cancel; "lb" in the numerator cancels "lb" in the denominator. The ratio is a dimensionless quantity, i.e., there are no units or dimensions.

If we were to compute the ratio of vehicle to pigment, the fraction would be inverted with vehicle on top and pigment on the bottom.

$$\frac{6\,\cancel{\text{lb}}}{12\,\cancel{\text{lb}}} = \frac{1}{2} \text{ or } 0.5$$

The ratio of A to B is always:

$$\frac{A}{B}$$

This ratio can also be written with a colon, A:B.

If the units do not cancel (disappear), the comparison is called a rate. If we compare the 10 miles we walked with the 2 hours it took us, we get:

$$\frac{\overset{5}{\cancel{10}}\text{ mi}}{\underset{1}{\cancel{2}}\text{ h}} = \frac{5\text{ mi}}{1\text{ h}} \text{ or } 5\text{ mi/h}$$

as our rate in mi/h.

We could also compare the 2 hours to the 10 miles by forming the reciprocal:

$$\frac{\overset{1}{\cancel{2}}\text{ h}}{\underset{5}{\cancel{10}}\text{ mi}} = \frac{1\text{ h}}{5\text{ mi}} = 0.2\text{ h/mi}$$

The rate is now expressed as 0.2 h/mi. Both of these rates tell us the same information, as we can always get one from the other.

The volatile organic compound (VOC) content of paint is a rate expressed as pounds per gallon (lb/gal) or grams per liter (g/L). If a paint has a VOC content of 500 g/L, then 500 g of VOCs are released from every liter of paint. This ratio can be inverted to give 1 liter releases 500 grams and written as:

$$\frac{1\ \text{L}}{500\ \text{g}} = \left(\frac{1}{500}\right)\left(\frac{\text{L}}{\text{g}}\right) = 0.002\,\frac{\text{L}}{\text{g}}$$

which tells us that 0.002 L of paint releases 1 g of VOCs. ($1 \div 500 = 0.002$).

A proportion is the equality of two ratios. Using letters, a proportion can be written in the form:

$$\frac{a}{b} = \frac{c}{d}$$

This is often stated as, "a is to b as c is to d." Some older books write proportions using colons a:b :: c:d. The a and d are called the extremes of the proportion (they occur at each end) and b and c are called the means (they are in the middle).

A proportion is an equation where one fraction is equal to another fraction. This form of equation occurs often, especially in chemistry and health sciences. In order to solve an equation of this type, there can only be one letter that is not known. The unknown letter may appear in any of the four positions. Rather than solve each proportion using the basic principles of multiplying, dividing, adding, or subtracting on both sides of the equation by the same number, a few shortcuts will be developed.

- The product of the extremes is equal to the product of the means.

$$ad = bc$$

Another way of stating this is that we can cross multiply to simplify a proportion. The same result occurs if we multiply both sides of the proportion by bd. The b will cancel on the left side and the d will cancel on the right side.

$$\frac{a}{b} = \frac{c}{d}$$
$$\frac{a}{\cancel{b}} \cdot \cancel{b}d = \frac{c}{\cancel{d}} \cdot b\cancel{d}$$
OR
$$ad = bc$$

• The means in a proportion can be exchanged. This happens if we multiply both sides of the proportion by b/c.

$$\frac{a}{b} = \frac{c}{d}$$
$$\frac{a}{\cancel{b}} \cdot \frac{\cancel{b}}{c} = \frac{\cancel{c}}{d} \cdot \frac{b}{\cancel{c}}$$
OR
$$\frac{a}{c} = \frac{b}{d}$$

Several other shortcuts can be derived in a similar fashion.

• The extremes in a proportion can be exchanged.

$$\frac{a}{b} = \frac{c}{d} \text{ can be rewritten } \frac{d}{b} = \frac{c}{a}$$

• Exchange both the means and the extremes, i.e., combine the previous two steps.

$$\frac{a}{b} = \frac{c}{d} \text{ can be rewritten } \frac{d}{c} = \frac{b}{a}$$

• Reverse the equation. This is true for all equations, not necessarily proportions.

$$\frac{a}{b} = \frac{c}{d} \text{ can be rewritten } \frac{c}{d} = \frac{a}{b}$$

• Flip the equation, i.e., if two fractions are equal, their reciprocals are equal.

$$\frac{a}{b} = \frac{c}{d} \text{ can be rewritten } \frac{b}{a} = \frac{d}{c}$$

• Take a factor from one numerator and place it in the other denominator; or move a factor from a denominator to the opposite numerator. We do not have to exchange means or extremes, we can move only one of the numbers.

$$\frac{a}{b} = \frac{c}{d} \text{ can be rewritten}$$

$$\frac{ad}{b} = \frac{c}{1}$$

OR

$$\frac{a}{1} = \frac{bc}{d}$$

OR

$$\frac{1}{b} = \frac{c}{ad}$$

Examples of Mixing Ratios: To obtain proper curing of multi-component coatings, we must adhere to the proper mixing ratio. Mixing ratios are also specified for chemical cleaning compounds and inhibitors.

Example: The lab develops a two-component epoxy that requires 0.21 gal of part A with 0.65 gal of part B. How much part A is needed for 5 gal of part B?

Solution: The first step is to determine the type of problem, i.e., a proportion. We want the ratio of part A to part B to be the same in the large batch as in the small batch. We form two ratios (part A to part B) and set them equal. Since we do not know the amount of part A in the large batch, we assign a variable (a letter) to it. Let M = amount of part A needed for the large batch. Then:

ratio for large batch = ratio for small batch

$$\frac{M}{5\text{ gal}} = \frac{0.21\text{ gal}}{0.65\text{ gal}}$$

Multiply both sides of this equation by 5 gal and we get:

$$M = \frac{0.21\cancel{\text{gal}}}{0.65\cancel{\text{gal}}} \cdot 5\text{gal}$$

$$M = 1.6\text{gal}$$

We need 1.6 gal of part A for 5 gal of part B. Notice that when we multiplied both sides by 5 gal, the unit traveled with the number. The units (gal in this example) follow the same rules of algebra as numbers.

Example: How much of part A and part B would we need so that the volume of the mixed paint is 5.0 gal?
Solution: When solving any problem using algebra, assign letters to those quantities we wish we knew, but don't. In this problem we do not know either the amount of part A needed or the amount of part B needed, so we assign letters to them. Let:

M = amount of part A needed
N = amount of part B needed

We know the ratio of A to B is 0.21 to 0.65. Hence we can write the relationship as:

$$\frac{M}{N} = \frac{0.021}{0.065} = \frac{0.323}{1}$$

or

$$\frac{M}{N} = 0.323$$

Since we have two unknowns (letters), we must find another relationship between M and N. We need as many equations as we have letters; in this case two. Since the total volume of the mixed paint is 5.0 gal, we can write :

$$M + N = 5$$

We must use some method of algebraic manipulation to combine these two equations into a single equation with only one letter. The general substitution principle is to solve one of the equations for a letter, and then substitute the expression for that letter into the other equation. In this case, let us multiply both sides of the first equation by N to get:

$$M = 0.323\ N$$

If we substitute this into the second equation we get:

$$0.323\ N + N = 5$$

Combine like terms to get:

$$1.323\ N = 5$$

Divided both sides by 1.323 to solve for N.

$$\frac{1.323\ N}{1.323} = \frac{5}{1.323}$$

so

$$N = 3.8$$

We now plug this value of N into one of our equations and solve for M.

$$M + N = 5$$

and since N = 3.8

$$M + 3.8 = 5$$

Subtracting 3.8 from both sides gives

$$M = 5 - 3.8$$
$$M = 1.2$$

Thus, we need 1.2 gal of part A and 3.8 gal of part B to make a 5.0 gal kit.

Example: In order to kill mildew and clean a mildew stained surface, a 1:4 dilution with water of household bleach (sodium hypochlorite) is recommended. How do we get such a concentration?
Solution: We first must decide what the 1 and the 4 represent. This is a ratio of 1 part bleach to 4 parts of SOLUTION. One fourth of the final solution will be bleach, the other three fourths will be water (the diluent).

Suppose we want to prepare one gallon of the diluted solution. Since there are 4 quarts in one gallon, we pour 3 quarts of water into a gallon container and then add one quart of bleach to fill up the container.

Example: What dilution do we get if we add one quart of bleach to four quarts of water?
Solution: The volume of the solution will be 5 quarts. The ratio of bleach to solution is then 1 quart to five quarts or 1:5. This is a more diluted solution than the 1:4 solution in the previous example, and will be less effective in killing the mildew.

Chapter 3 – Dimensional Analysis

Objective: To review systems of measurement and learn how to convert from one unit (dimension) to another. To derive the conversion factors 1604 mil-ft^2 = 1 gal and 1000 $\mu m \cdot m^2$ = 1 L.

The dimensions or units written after a number convey meaning to the number. Writing 8 ft is not the same as 8 lb or 8 CFM. The number 8 by itself is a dimensionless quantity. The dimensions ft, lb, or CFM add meaning to the number. A man may list his weight as 260 lb or 118 kg. The same weight can be expressed with different units.

To develop a system of measurement units, we must first define some basic or fundamental quantities, such as mass, length, and time. In customary U. S. units, length is measured in feet and time is measured in seconds. Feet and seconds are fundamental units. Speed, on the other hand, is a derived unit, since we measure speed in feet per second. Once we have defined one foot and one second, we know how fast we are moving at a speed of one foot per second. Similarly, volume, which can be measured in cubic feet, is a derived unit. If we know one foot in length, we know the volume of a cube one foot on a side.

3.1 The Metric System

The metric system was developed by the French around 1800 AD. The "meter" was one ten-millionth of the distance from the North Pole to the equator along the quadrant through Paris. The distance between two scratches on a platinum-iridium (Pt-Ir) bar defined the meter. This "meter of the archives" served as the world standard of length until the mid-1900s when atomic standards were adopted.

The metric standard for mass was a Pt-Ir cylinder of a weight, one kilogram, equal to the weight of 0.001 cubic meters (one liter) of water. This metallic cylinder still serves as the world standard of mass. Since the mid-1900s, the standard unit for time, the second, has been defined by an atomic clock.

The standard of length for the U. S. is the meter. The inch is defined as 0.0254 meters, exactly. Similarly, the pound mass is defined in terms of the standard kilogram. The second is the international unit for time. The National Institute for Standards and Technology (NIST) in Gaithersburg, MD, has a Pt-Ir cylinder with a mass of one kilogram, which is the standard for mass in the United States. NIST also has the equipment that measures a standard meter, a standard second, and other standard quantities.

There are many standard metric units derived from the fundamental units. The metric unit for volume is the liter (L); 1 L = 0.001 m^3 or 1000 cm^3 or 1000 cc. A newton is the unit of force, 1 N = 1 kg m/s^2. The standard unit for pressure is the pascal, 1 Pa = 1 N/m^2.

The modern metric system is called the International System of Units (or SI units). The "S" comes before the "I" because in the French language the adjective follows the noun.

Multiples for the metric quantities are based on powers of 10. The set of metric prefixes determines how many times we multiply or divide by 10 to change the units from one metric quantity to another metric quantity. Use of the metric system and the complete list of prefixes is given in IEEE/ ASTM SI 10, Standard for Use of the International System of Units (SI):The Modern Metric System (Replaces ASTM E 380 and ANSI/IEEE Std 268-1992). Chemistry, physics, and applied math books also contain tables of metric prefixes and derived units. A few of the common metric prefixes are given Tables 3.1 and 3.2. Multiples smaller than the base unit are given in Table 3.1 and multiples larger than the base unit are given in Table 3.2.

Table 3.1

Some metric prefixes for multiples smaller than the base unit

PREFIX	MULTIPLIER	POWER OF 10	ABBREVIATION
deci *	0.1	10^{-1}	d
centi *	0.01	10^{-2}	c
milli	0.001	10^{-3}	m
micro	0.000 001	10^{-6}	μ
nano	0.000 000 001	10^{-9}	n

* These prefixes are to be avoided where practical.

Table 3.2

Some metric prefixes for multiples larger than the base unit

PREFIX	MULTIPLIER	POWER OF 10	ABBREVIATION
deka *	10	10^1	da
hecto *	100	10^2	h
kilo	1,000	10^3	k
mega	1,000,000	10^6	M
giga	1,000,000,000	10^9	G

* These prefixes are to be avoided where practical.

A few examples to illustrate use of the metric prefixes:

1 mm = 0.001 m — 1 millimeter is equal to 0.001 meter — 1000 mm = 1 m
1 μS = 0.000 001 S — 1 microsiemens is equal to 0.000 001 siemens — 1,000,000 μS = 1 S
1 kPa = 1000 Pa — 1 kilopascal is equal to 1000 pascals

The official ASTM guideline is to avoid prefixes deci, centi, deka, and hecto, but they have been ingrained in many scientific fields for a long time. Every chemist knows 1cc (1 cm^3). A blood test for cholesterol may show 250 mg/dL.

3.2 Conversion of Units

One of the most useful skills that a scientist, engineer, or worker dealing with numerical quantities can have is the ability to understand the conversion from one unit to another. For example, our equipment may be rated in CFM but the governmental regulations are in L/min or m^3/min. We must convert to the same system before we can test for compliance.

There are different methods of approach to the conversion of units problem. One method is to find a table that says, for example, "To convert m to mm, multiply by 1000." Tables such as this are common in government and other standard publications. Computer software is available that will convert units. However, if we do not have access to the table with our particular conversion and we do not have the computer software, or if the conversion is sufficiently critical to warrant verification, we must resort to some variation of the method presented here.

We will start with the simple problem of converting 24 ounces to pounds. We may be able to figure out the answer quickly in our heads, but this example is to illustrate a method. Before starting, we must know or look up the equivalence between the two quantities of interest, 16 oz = 1 lb. We can form two fractions, each equal to 1, as shown:

$$\frac{16\text{ oz}}{1\text{ lb}} = 1$$

$$\frac{1\text{ lb}}{16\text{ oz}} = 1$$

Since the numerator is equal to the denominator in each of these fractions, the fractions are equal to 1. We can always multiply by 1 and not change the value. The problem is to decide which of the above fractions, each equal to 1, we should use. We begin by writing the quantity we wish to convert as a fraction with a denominator of 1. Any number can be written as a fraction with a denominator of 1. Then we multiply by the fraction equal to 1 such that the units cancel.

$$\frac{24\,\cancel{\text{oz}}}{1} \cdot \frac{1\text{ lb}}{16\,\cancel{\text{oz}}} = 1.5\text{ lb}$$

$$24\text{ oz} \cdot 1 = 1.5\text{ lb}$$

The 1.5 comes from dividing 24 by 16 on a calculator. The "oz" cancel leaving "lb" as the remaining unit. If the units do not cancel, then we used the wrong form of the fraction.

Example: Ultrahigh pressure waterjetting (UHP WJ) uses pressures above 35,000 pounds per square inch (psi). Convert 35,000 psi to kilopascals (kPa) and megapascals (MPa).
Solution: We must first use a table (such as ASTM SI 10 or the tables in Appendix B of this book) to find that 1 psi = 6.895 kPa. We write the quantity we know as a fraction with a denominator of 1. We multiply by a fraction equal to 1 with psi in the denominator so that the psi will cancel.

$$\frac{35{,}000\,\cancel{\text{psi}}}{1} \cdot \frac{6.895\text{ kPa}}{1\,\cancel{\text{psi}}} = 240{,}000\text{ kPa}$$

Since MPa and kPa are both metric units, we need only move the decimal point to convert from one to the other. From Table 3.2 we find that kilo means times 1000 and mega means times 1,000,000; so 1 MPa is 1000 kPa. We can use reasoning to determine that we will not have as many MPa as kPa so we must move the decimal point three places to the left. Or we can repeat the process used above.

One method is to convert to the base unit and then to the unit we want; in this case we convert kPa to Pa and then convert from Pa to MPa.

$$\frac{240{,}000\ \cancel{\text{kPa}}}{1} \cdot \frac{1000\ \cancel{\text{Pa}}}{1\ \cancel{\text{kPa}}} \cdot \frac{1\ \text{MPa}}{1{,}000{,}000\ \cancel{\text{Pa}}} = 240\ \text{MPa}$$

Or, if we are familiar with the metric prefixes, we can convert directly from kilopascals to megapascals.

$$\frac{240{,}000\ \cancel{\text{kPa}}}{1} \cdot \frac{1\ \text{MPa}}{1000\ \cancel{\text{kPa}}} = 240\ \text{MPa}$$

Note that each of the fractions that we multiply by is equal to 1 because the numerator of the fraction is equal to the denominator. These fractions are called conversion factors. A factor is something we multiply by, and the result is to convert the units.

Example: Suppose we want to convert 9 in² to cm² but do not have a table handy. However we do know that 1 in = 2.54 cm, i.e., we know the length conversion but not the area conversion.
Solution: We simply use the length conversion factor twice.

$$\frac{9\ \cancel{\text{in}^2}}{1} \cdot \frac{2.54\ \text{cm}}{1\ \cancel{\text{in}}} \cdot \frac{2.54\ \text{cm}}{1\ \cancel{\text{in}}} = 58\ \text{cm}^2$$

We must use the "in" factor twice in the denominator because the numerator has in². The numerical computation is done on a calculator.

Sometimes we want to convert units in both the numerator and the denominator. It is possible to use several different factors in the same conversion.

Example: Convert 60 mi/h to ft/s.
Solution: Assume we do not know the factor that takes us directly from

mi/h to ft/s. However, we do know 5280 ft = 1 mi, 1 h = 60 min, and 1 min = 60 s. We proceed by converting mi to ft, h to min, and min to sec.

$$\frac{60\ \text{mi}}{1\ \text{h}} \cdot \frac{5280\ \text{ft}}{1\ \text{mi}} \cdot \frac{1\ \text{h}}{60\ \text{min}} \cdot \frac{1\ \text{min}}{60\ \text{s}} = \frac{88\ \text{ft}}{1\ \text{s}}$$

It is not necessary to write the 1 in the denominators, but it may help in viewing each number as a fraction. The result is that 60 mi/h = 88 ft/s.

When using metric conversion factors for smaller multiples of the base unit, it is often convenient to eliminate the decimal before doing the computation. Although 1 mm = 0.001 m, it is easier to think of 1000 mm = 1 m. This can be done if we write the conversion as a fraction and multiply the numerator and the denominator by the appropriate power of 10.

$$\frac{1\ \text{mm}}{0.001\ \text{m}} \cdot \frac{1000}{1000} = \frac{1000\ \text{mm}}{1\ \text{m}}$$

Often we may find ourselves converting between units that are not on our list of conversion factors. It is then useful to derive our own conversion factor so that we can make subsequent conversions in a single step.

Example: Environmental requirements for volatile organic compound (VOC) content of coatings are given in pounds per gallon or grams per liter. Using the common conversions 1 lb = 453.6 g and 1 gal = 3.785 L, find the conversion factors from lb/gal to g/L and from g/L to lb/gal.
Solution: Let's start by converting 1 lb/gal to g/L. This involves converting the pounds to grams and the gallons to liters. We will multiply by two conversion factors (fractions) each equal to one. In order for the lb unit to cancel, the first factor will have lb in the denominator. To cancel the gal, the conversion factor must have gal in the numerator.

$$\frac{1\ \text{lb}}{1\ \text{gal}} \cdot \frac{453.6\ \text{g}}{1\ \text{lb}} \cdot \frac{1\ \text{gal}}{3.785\ \text{L}} = 119.8\ \frac{\text{g}}{\text{L}}$$

If we wish, we can round this off to give 1 lb/gal ≈ 120 g/L.

We perform a similar operation when we derive the conversion from g/L to lb/gal. Starting with 1 g/L we multiply by conversion factors oriented so that the units cancel.

$$\frac{1\,\cancel{g}}{1\,\cancel{L}} \cdot \frac{1\text{ lb}}{453.6\,\cancel{g}} \cdot \frac{3.785\,\cancel{L}}{1\text{ gal}} = 0.008344\,\frac{\text{lb}}{\text{gal}}$$

If we wish, we can round this to 1 g/L = 0.0083 lb/gal.

Either of the conversions could have been obtained from the other by taking the reciprocal (inverting the fraction). For example, once we found that 1 lb/gal = 119.8 g/L, we could have divided both sides of the equation by 119.8 (or multiplied by 1/119.8) to get:

$$\frac{119.8\text{ g}}{1\text{ L}} = \frac{1\text{ lb}}{1\text{ gal}}$$

$$\frac{119.8\text{ g}}{1\text{ L}} \cdot \frac{1}{119.8} = \frac{1\text{ lb}}{1\text{ gal}} \cdot \frac{1}{119.8}$$

$$\frac{1\text{ g}}{1\text{ L}} = \frac{1\text{ lb}}{1\text{ gal}} \cdot 0.008347$$

OR

$$1\frac{\text{g}}{\text{L}} = 0.008347\,\frac{\text{lb}}{\text{gal}}$$

With a hand-held calculator, all we need to do is enter 119.8 and hit the "1/x" button.

$$\frac{1}{119.8} = 0.008347 \quad \text{and} \quad \frac{1}{0.008347} = 119.8$$

If we were to use all the significant figures given in ASTM SI 10, the conversion factors would be (to four significant figures):

$$1\frac{\text{lb}}{\text{gal}} = 119.8\frac{\text{g}}{\text{L}} \quad \text{and} \quad 1\frac{\text{g}}{\text{L}} = 0.008345\frac{\text{lb}}{\text{gal}}$$

Example: Convert 2.2 lb/gal to g/L.
Solution: When converting compound units in one step, we can treat the compound unit as a single unit.

$$\frac{2.2\,\cancel{(\text{lb/gal})}}{1} \cdot \frac{119.8(\text{g/L})}{1\,\cancel{(\text{lb/gal})}} = 263.6(\text{g/L})$$

After rounding we find that 2.2 lb/gal is equal to 260 g/L.

3.3 Deriving Paint Coverage Using Unit Conversions

U. S. Customary Units: It is known among those in the paint industry that the theoretical coverage for a gallon of paint applied at a wet film thickness of 1 mil is 1604 square feet. The metric thinkers know that one liter of paint applied at a wet film thickness of one micrometer will cover 1000 square meters. This section will explain how these numbers come from a conversion of units.

Imagine a rectangular box with base area [A] and height [H]. The volume [V] of the box is then $V = A \cdot H$, or omitting the multiplication sign, $V = AH$. The more familiar formula for the volume of a box is length times width times height, but length times width is the area of the base. Now imagine the box to be flattened out so that the height is a very small number. The volume is still $V = AH$. This flattened box can be the coat of paint.

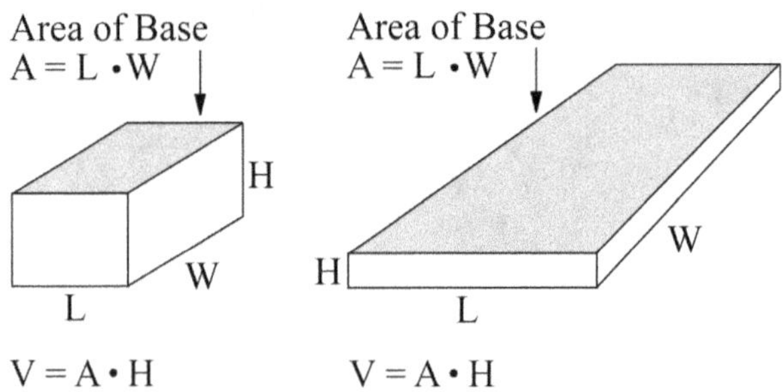

We begin by letting the volume of the box (the paint volume) be one gallon. Let the height of the box (the paint thickness) be one mil. We must now solve for the area of the base of the box (the area covered by the paint).

If we rearrange the volume formula by dividing both sides of the equation by H we get the formula for area:

$$V = A \cdot H$$

$$\frac{V}{H} = \frac{A \cdot H}{H}$$

$$\frac{V}{H} = A$$

Reversing the equation gives:

$$A = \frac{V}{H}$$

We know the volume in gallons and the height in mils and we want to find the area in square feet. We need to know the following equivalents, which can be found in a table (Appendix B):

$$1 \text{ gal} = 231.0 \text{ in}^3$$
$$1 \text{ mil} = 0.001 \text{ in}$$
$$1 \text{ ft}^2 = 144 \text{ in}^2$$

We substitute V = 1 gal and H = 1 mil into the formula and then multiply by the appropriate conversion factors.

$$A = \frac{1 \text{ gal}}{1 \text{ mil}} \cdot \frac{231.0 \text{ in}^3}{1 \text{ gal}} \cdot \frac{1 \text{ mil}}{0.001 \text{ in}} \cdot \frac{1 \text{ ft}^2}{144 \text{ in}^2} = 1604 \text{ ft}^2$$

The gal, mil, and inches all cancel leaving ft^2 in the numerator. Thus, if the volume is 1 gal and the thickness is 1 mil, the area is 1604 ft^2. In most instances, three significant figures are adequate and this can be rounded to 1600 ft^2.

Since A = V/H, if we double the H, we get half the area; if we triple the H, we get one third the area; and so on. So one gallon applied at 2 mils will cover 802 ft^2 [1604 ÷ 2 = 802 ft^2].

SI (Metric) Units: Repeating the process above using a volume of 1 L and a thickness of 1 μm, we compute the area in square meters. We need to know 1 L = 0.001 m^3 and 1μm = 0.000 001 m.

$$A = \frac{1 \text{ L}}{1 \,\mu\text{m}} \cdot \frac{0.001 \text{ m}^3}{1 \text{ L}} \cdot \frac{1 \,\mu\text{m}}{0.000001 \text{ m}} = 1000 \text{ m}^2$$

We could have avoided the decimals by using the conversions 1000 L = 1 m^3 and 1,000,000 μm = 1 m.

$$A = \frac{1\,\cancel{L}}{1\,\cancel{\mu m}} \cdot \frac{1\,\cancel{m^3}^{\,m^2}}{1000\,\cancel{L}} \cdot \frac{1000000\,\cancel{\mu m}}{1\,\cancel{m}} = 1000\ m^2$$

Notice with metric conversions all we have to do is move the decimal point.

3.4 Temperature Conversions

The two most common temperature scales are the Fahrenheit scale and the Celsius scale. The Celsius scale has also been called the centigrade scale. Immerse two identical thermometers without any markings on them into ice water. Mark the mercury level. Label this mark 32 on the first thermometer and 0 on the other one. Next immerse the two thermometers in boiling water and mark the mercury level. Label this mark 212 on the first thermometer and 100 on the second one.

The first thermometer has 180 units (212 – 32 = 180) between the freezing point and the boiling point of water. If we mark 180 equally spaced divisions between the initial two marks, we have the Fahrenheit temperature scale. We can continue marking these divisions above 212 and below 32.

On the second thermometer we make 100 equally spaced divisions between the two initial marks, and then continue above 100 and below 0. This is the Celsius scale.

Comparing these two scales we see that a change of 180 Fahrenheit degrees is equivalent to a change of 100 Celsius degrees. If we reduce the fraction 180/100 to lowest terms we get 9/5. So a change of 9 Fahrenheit degrees is equal to a change of 5 Celsius degrees. Expressing 180/100 as a decimal gives 1.8, so it takes 1.8 Fahrenheit degrees to make one Celsius degree.

Example: During painting operations, we want the surface temperature to be 5 Fahrenheit degrees above the dew point. How many Celsius degrees above the dew point is this?

Solution: In this type of conversion we are not comparing the reading on two different thermometers. Those examples will come later in this chapter.

Figure 1. Magnetic Surface Temperature Thermometer. Courtesy Elcometer, Inc.

Figure 2. Digital Thermometer. Courtesy Elcometer, Inc.

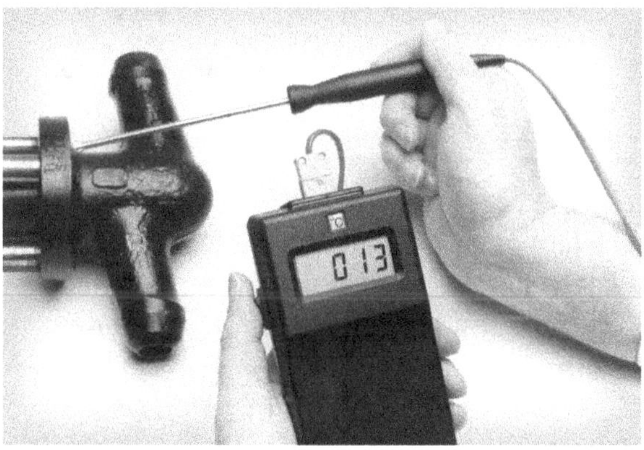

Here, we are comparing a temperature CHANGE or interval. Each Celsius degree is 1.8 Fahrenheit degrees. Doing the conversion for temperature intervals we get:

$$\frac{5\ \cancel{\text{Farenheit}}}{1} \cdot \frac{1\ \text{Celsius}}{1.8\ \cancel{\text{Farenheit}}} = 2.8\ \text{Celsius}$$

The 2.8 can be rounded to 3. Thus, good painting practice requires the surface temperature to be 5 Fahrenheit degrees or 3 Celsius degrees above the dew point.

A different question is to compare the temperature measured on the

Fahrenheit scale with the temperature measured on the Celsius scale. We know from the way these scales were determined that 0°C = 32°F and 100°C = 212°F. Since both scales are linear (the degrees are equally spaced), we can use these two reference points to derive the relationship between the scale readings. The resulting equations relating these temperature scales are given below.

Formulas for Conversion Between Fahrenheit and Celsius Temperatures

☞ If we know F and want to find C, then we use

$$C = \frac{F - 32}{1.8}$$

☞ If we know C and want to find F, then we use

$$F = 1.8C + 32$$

where:

F = the temperature reading in degrees Fahrenheit
C = the temperature reading in degrees Celsius

Example: The heat-resisting paint TT-P-28 is effective at temperatures up to 650°C. What is this temperature in degrees Fahrenheit?
Solution: Since we know C and want to find F, we use:

$$F = 1.8\,C + 32$$
$$F = 1.8\,(650) + 32$$
$$F = 1170 + 32$$
$$F = 1202$$

This can be rounded to 1200°F.

Example: SSPC-PA 1 recommends a minimum temperature of 40°F for storing latex paints. What is this temperature on the Celsius scale?

Solution: We know F and wish to find C, so we use the first formula.

$$C = \frac{F - 32}{1.8}$$
$$C = \frac{40 - 32}{1.8}$$
$$C = \frac{8}{1.8}$$
$$C = 4.4$$

This can be rounded to 4°C as the minimum storage temperature.

Example: Our friend from Canada tells us they were painting at 10 degrees below zero (–10°C). What is this temperature in degrees Fahrenheit?
Solution: We know C and want to find F, so we use the second formula.

$$F = 1.8\,C + 32$$
$$F = 1.8\,(-10) + 32$$
$$F = -18 + 32$$
$$F = 14$$

Thus –10°C corresponds to 14°F.

Absolute temperature scale: The coldest temperature that can theoretically be achieved is –273.15°C (–459.17°F). This temperature is called absolute zero. Absolute zero is usually rounded to –273°C or –459°F.

If we re-label a Celsius thermometer so that the coldest temperature is zero, we will have all positive temperatures. We keep the size of each degree the same as a Celsius degree. This is the absolute temperature scale. The unit of temperature is kelvin (K).The degree symbol is not used when expressing temperature measurements on this scale.

Absolute zero is then written as 0 K. Water freezes at 0°C, which is 273 degrees above absolute zero; hence, 0°C = 273 K. Similarly, water boils at 100°C or 373 K.

To convert from Celsius temperature, C, to kelvins, K, add 273 to the Celsius temperature.

$$K = C + 273$$

Subtract 273 from the absolute temperature, K, to get the Celsius temperature, C.

$$C = K - 273$$

If temperatures are given in degrees Fahrenheit, convert to degrees Celsius and then to kelvins.

There are several computer programs that will compute unit conversions and can be downloaded for free from the Internet. Two sites where these programs are available are:

- http://www.esbconsult.com/esbcalc/esbunitconv.html
- http://www.joshmadison.com/software/convert/

Chapter 4 – Geometry

Objective: To review the formulas for computing the area of basic shapes and the volume of basic solid objects. To define weight and mass density. To develop an understanding of the relationship among weight, mass, volume, density, and specific gravity through applications.

4.1 Area

Before submitting a bid for a job, it is necessary to know the area of the structure to be covered. The areas of more complex shapes or structures are computed by first breaking down complex structures into simple structures.

The basic unit of area is the square. If the side of the square is 1 ft, the area of the square is one square foot (1 ft^2). Scientists write square foot with the exponent, ft^2, because it is easier to convert from one measurement system to another and it is easier to do dimensional analysis in general. If the square has a side of one meter, the area is one square meter (1 m^2). The area of any shape (e.g., circle, triangle, ellipse) can be measured in ft^2, m^2, etc.

Square: The area of a square (each side of length s) is the length of a side times itself, or the side squared.

$$A = s^2$$

Example: Find the area of a square 3 inches on a side.

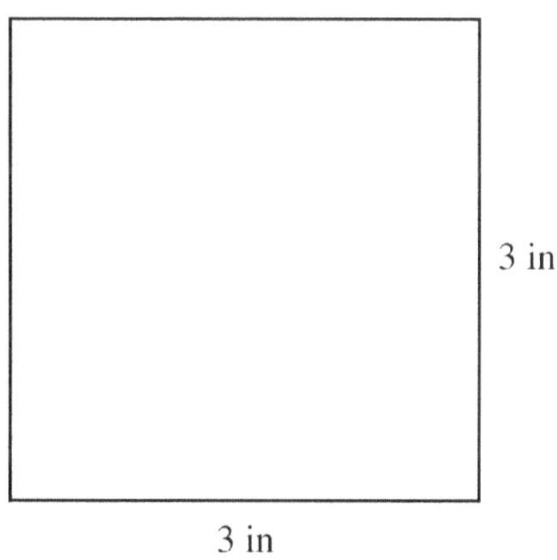

Solution: Substitute 3 inches for s in the area formula.

$$A = (3\text{ in})^2$$
$$A = (3\text{ in})(3\text{ in})$$
$$A = 9\text{ in}^2$$

Rectangle: The area of a 2 m x 3 m rectangle is 6 m^2 as shown in the figure below. Each block represents an area of 1 m^2 since the side of each square block is 1 m. Six of these blocks fit inside the rectangle.

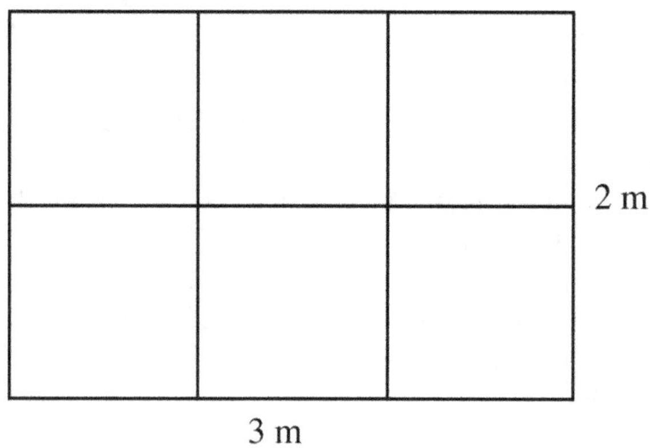

The formula for the area of a rectangle is:

$$A = L \cdot W$$

where L is the length and W is the width.

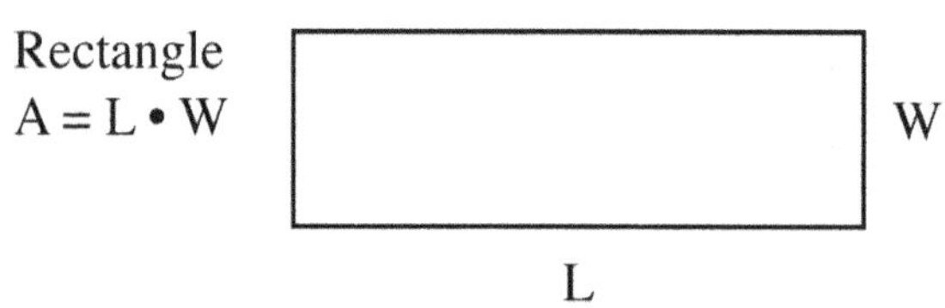

Example: Find the area of the L-shaped figure below.

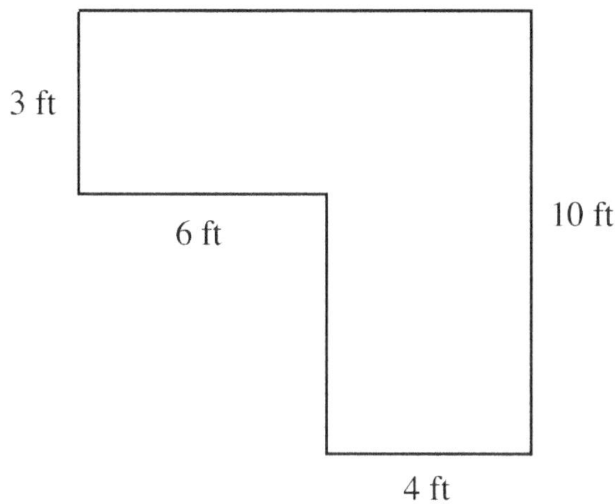

Solution: The first step is to divide the L-shaped figure into two rectangles as shown.

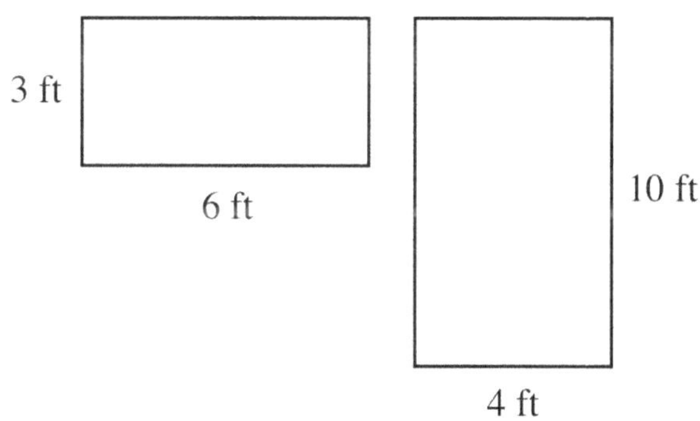

The area of the first rectangle is 3 ft x 6 ft = 18 ft^2; the area of the second rectangle is 10 ft x 4 ft = 40 ft^2. So the total area of the L-shaped figure is 18 ft^2 + 40 ft^2 = 58 ft^2. Other area formulas are given on the pages that follow.

Triangle

area = 1/2 base x height

$$A = \frac{1}{2} b \cdot h$$

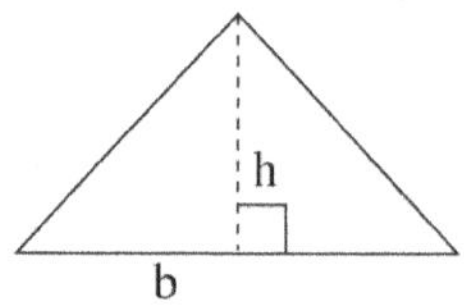

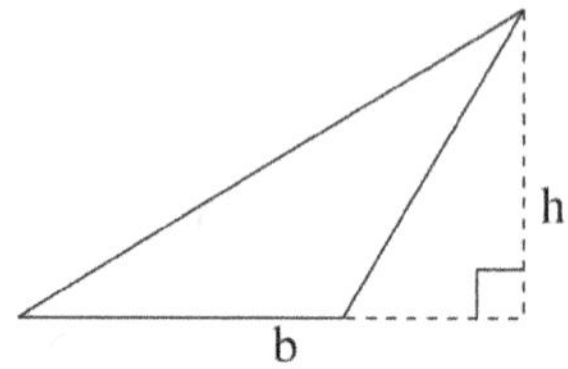

Choose one side of a triangle as the base. The height is then the distance from the opposite vertex (point of the angle) to the line of the base (which may have to be extended).

Parallelogram

area = base x height
$A = b \cdot h$

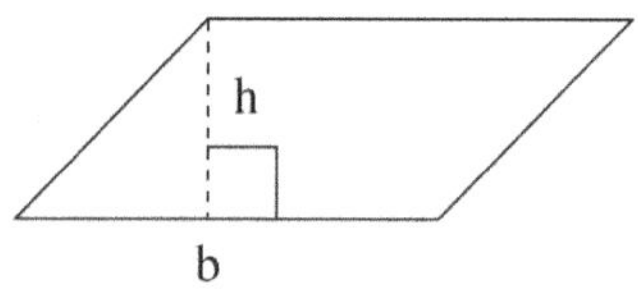

Trapezoid

area = average base x height

$$A = \frac{B + b}{2} \cdot h$$

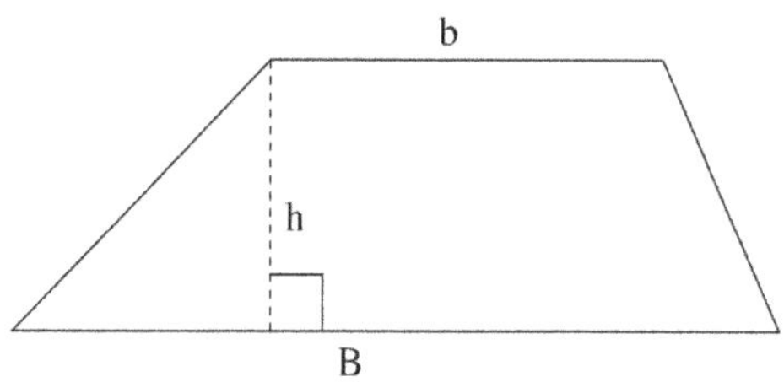

Perimeter: Another quantity of interest when describing closed figures like squares and rectangles is the perimeter, or distance around the outside. Unlike area, which is measured in square feet or square meters, perimeter is a length and is measured in linear dimensions such as feet or meters.

For a quadrilateral, a closed figure with four sides, the perimeter is the sum of the four sides. Since all sides of a square are equal, the perimeter formula is $P = 4s$. In a rectangle, two pairs of sides are equal, so the perimeter formula is:

P = 2L + 2W

where

P = perimeter
L = length
W = width

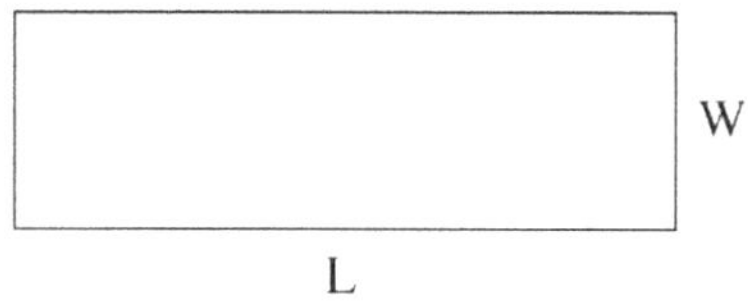

Circles: Quantities of interest when describing a circle are radius, diameter, circumference, and area.

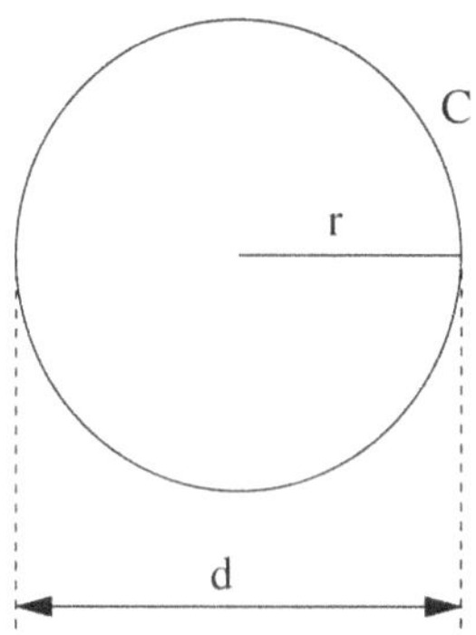

The **radius, r,** is the distance from the center to the edge of the circle. The **diameter, d,** is the distance from one edge through the center to the other edge. The diameter is twice the radius.

The **circumference, C,** is the distance around the outside of the circle, i.e., the perimeter of the circle. In ancient times, it was discovered that the ratio of the circumference to the diameter was the same for any circle. Numerically, C/d = 3.14159... This number was given the name pi and the symbol π. Pi is not a rational number, which means it cannot be expressed as an exact fraction or decimal.

Pi is an irrational number like the square root of 2, which can be approximated to as many decimal places as we want when performing calculations. Most hand-held calculators have a "π" button that will enter all the digits with a single touch. If the calculator does not have this button, π can be approximated by 3.14 or 22/7.

The formula for circumference of a circle is then:

$$C = \pi d$$

or in terms of radius

$$C = 2\pi r$$

The area of the circle can be calculated from the formula

$$A = \pi r^2$$

or if we replace r with d/2, the formula for area becomes

$$A = \pi \cdot \left[\frac{d}{2}\right]^2 = \pi \frac{d^2}{4}$$

Notice that the area formula contains a length squared, so that the dimensions of area are square units (ft^2, m^2, etc.). Circumference, diameter, and radius are lengths and are therefore measured in linear units such as feet or meters.

Example: The circumference of a closed cylindrical tank with a flat roof is measured to be 138 ft, and the height is 15.0 ft. What is the radius of the tank and what is the area of the roof? What is the area of the side wall?
Solution: We know the circumference, C, and we are looking for the radius, r. The formula that relates what we know with what we do not know is $C = 2\pi r$. Substituting the known quantities into the formula we get:

$$138 \text{ ft} = 2\pi r$$

If we divide both sides by 2π we get: $\frac{138 \text{ ft}}{2\pi} = r$

Doing the computation on a calculator and rounding, we find that $r = 22.0$ ft. Note that π is a dimensionless quantity and has no units.

Once we know the radius we can compute the area of the roof from the formula $A = \pi r^2$.

$$A = \pi(22.0 \text{ ft})^2 = 1520.53 \text{ ft}^2$$

which can be rounded to

$$A = 1520 \text{ ft}^2$$

To compute the area of the side wall of the tank, consider the side to be made up of a flat rectangular sheet bent into a circle. The length of the flat sheet before bending into the circle is the circumference of the tank. Hence, the area of the side wall is the circumference times the height.

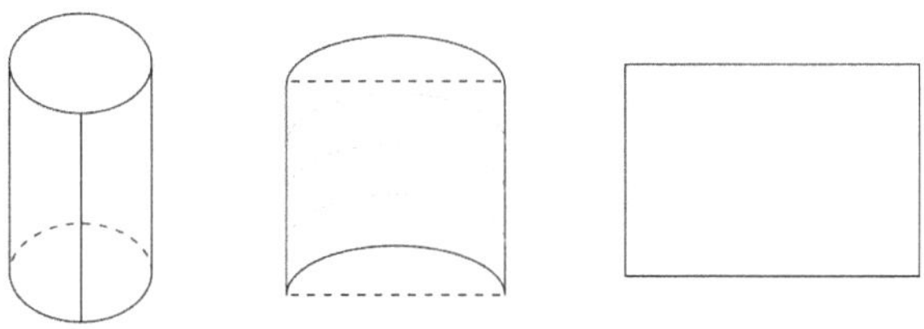

$$A_{side} = (138 \text{ ft})(15.0 \text{ ft}) = 2070 \text{ ft}^2$$

The total surface area of the tank exterior is then the area of the side wall plus the area of the top, or 3590 ft^2. [2070 ft^2 + 1520 ft^2 = 3590 ft^2].

In practice, any ladders, structural framing and support, or other protuberances must be accounted for when preparing a bid. When painting the interior, the area of the floor and underside of the roof must also be considered.

Cylinder: For a cylinder of radius, r, and height, h, the area of the roof (or floor) is the area of a circle, namely, $A_{Roof} = \pi \cdot r^2$. The area of the side wall, $A_{Side} = 2\pi \cdot r \cdot h$, can also be thought of as the circumference times the height.

When computing the area of a regular object, such as a pipe, the area is the circumference times the length. We can use reasoning similar to what we used in finding the area of the tank side wall in the previous example.

Example: An H-beam has a perimeter of 40 inches and is 20 ft long. What is the area, excluding the ends? What is the area of each end?

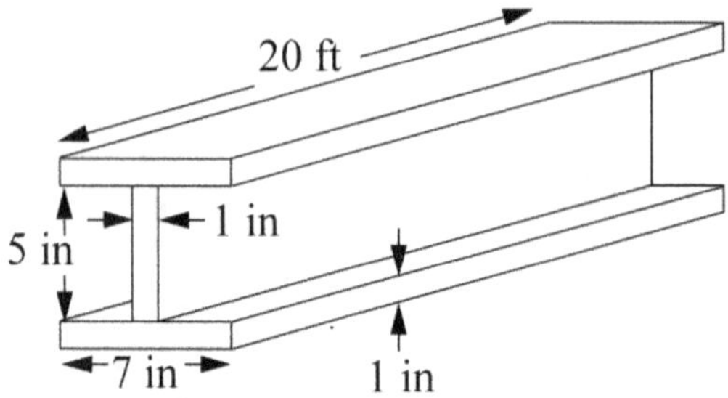

Solution: The perimeter is measured in inches, so we first convert it to feet.

$$\frac{40\ \cancel{in}}{1} \cdot \frac{1\ ft}{12\ \cancel{in}} = 3.333\ ft$$

It is good practice to carry more significant figures in the intermediate calculations and then round the final answer. Next we substitute into the formula:

$$A = P \cdot L$$
$$A = (3.333\ ft)(20\ ft)$$
$$A = 66.667\ ft^2 \text{ which can be rounded to } 67\ ft^2.$$

The area of one end is found by dividing the end into three rectangles. The top and bottom rectangles each have an area of 7 in^2 and the web cross section is 5 in^2 for a total cross sectional area of $7 + 7 + 5 = 19\ in^2$. The other end will also have an area of 19 in^2.

If we include the ends in our area calculation, we will add 38 in^2. How does this compare to 67 ft^2?

General Area Formula

☞ The surface area (excluding the ends) of a regular object of length, L, and perimeter, P, is

$$A = P \cdot L$$

The cross section of the piece must be uniform so that the perimeter is the same at all points along the piece.

Convert 38 in² to square feet.

$$\frac{38\ \text{in}^2}{1} \cdot \frac{1\ \text{ft}^2}{144\ \text{in}^2} = 0.26\ \text{ft}^2$$

Thus, the area of the ends (0.26 ft²) is negligible compared to the area of the sides of the beam (67 ft²).

Example: A pipe has an internal diameter of 2.0 ft and is 5.3 miles long. What is the internal circumference of the pipe? What is the interior surface area of the pipe?
Solution: The circumference of the pipe interior is:

$$C = \pi d = \pi\,(2.0\ \text{ft}) = 6.28\ \text{ft}$$

In computing the area, we can ignore the bends in the pipe and consider it a regular (straight) object. The area is then the perimeter (circumference) of the pipe times its length. We must also include the conversion factor from miles to feet. Substituting into the area formula we get:

$$A = (\pi d) \cdot L = (\pi \cdot 2.0\ \text{ft}) \cdot \left[\frac{5.3\ \text{mi}}{1} \cdot \frac{5280\ \text{ft}}{1\ \text{mi}}\right] = 175{,}829\ \text{ft}^2$$

which we round to 180,000 ft².

4.2 Volume

A cube with an edge length of 1 ft is said to occupy a volume of one cubic foot (1 ft³). Similarly, if the edge length is 1 m, the volume will be 1 m³. As mentioned in Section 3.1 of this book, a volume of 0.001 m³ is defined as one liter (1 L). Thus, 1000 L = 1 m³. Using the metric prefixes, a milliliter is 0.001 liter or 1000 mL = 1 L. If we convert 0.001 m³ into cm³ we get:

$$1\ \text{L} = \frac{0.001\ \text{m}^3}{1} \cdot \left(\frac{100\ \text{cm}}{1\ \text{m}}\right)^3 = 1000\ \text{cm}^3$$

Since 1 L = 1000 mL = 1000 cm³, we see that 1mL = 1 cm³. In chemistry and the health sciences, one cubic centimeter is abbreviated as 1 cc. Thus:

$$1\ \text{mL} = 1\ \text{cm}^3 = 1\ \text{cc}$$

For a rectangular solid, the area of the base is length, L, times width, W; so if the height is H, the volume formula is:

$$V = L \cdot W \cdot H$$

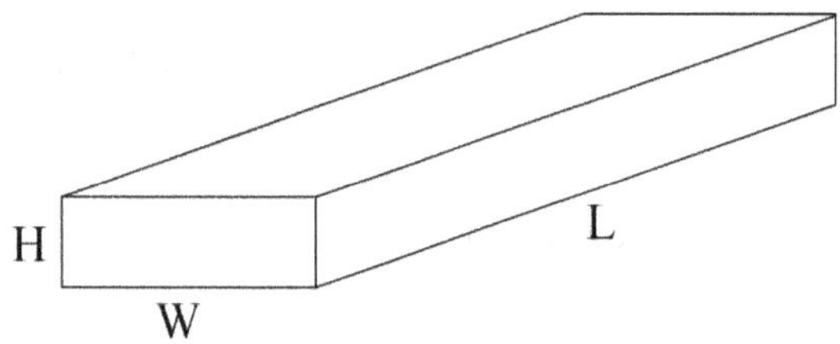

Example: What is the volume of a box 2.0 ft long, 1.5 ft wide, and 9.0 inches deep?

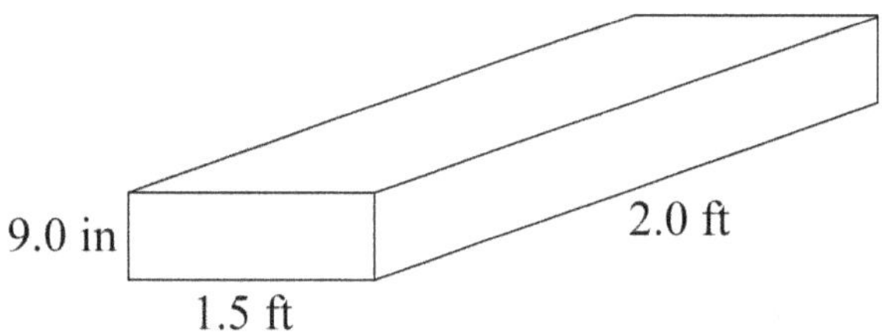

Solution: We first identify the known quantities L, W, and H. Notice that to get volume in cubic feet, we must have L, W, and H in feet. Before these are substituted into the formula for volume, let us convert 9 inches to feet:

$$\frac{9 \not{in}}{1} \cdot \frac{1 \text{ ft}}{12 \not{in}} = 0.75 \text{ ft}$$

General Volume Formula

☞ The volume of a regular solid (e.g., a block or a cylinder) is computed by multiplying the area of the base times the height.

$V = (\text{area of base}) \times (\text{height})$

Substituting

$$V = L \cdot W \cdot H$$
$$V = (2.0\ ft)(1.5\ ft)(0.75ft)$$
$$V = 2.25\ ft^3$$

which can be rounded to 2.2 ft^3.

The volume of a cylinder is the area of the base times the height. For a cylinder of radius, r, the base is a circle with area, πr^2. Multiplying this area by the height, h, we get the volume:

$$V = \pi r^2 h$$

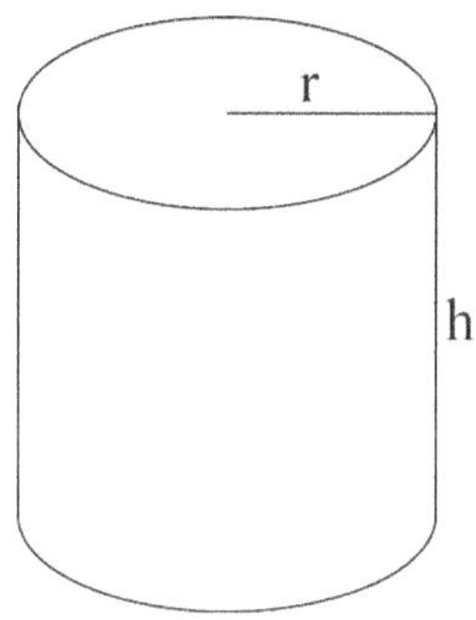

Example: What is the volume of a cylindrical tank with a radius of 22.0 ft and a height of 15.0 ft? How many gallons will it hold? How many liters?

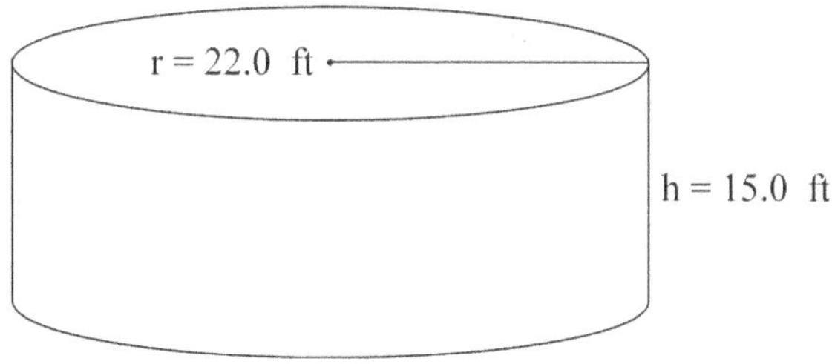

Solution: We know the radius and the height in feet, so we can compute the volume in cubic feet by substituting into the formula:

$$V = \pi r^2 h$$

$$V = \pi(22.0\ \text{ft})^2(15.0\ \text{ft})$$

$$V = 22{,}808\ \text{ft}^3$$

After rounding we find the volume of the tank to be 22,800 ft^3.

To convert to gallons we look up the conversion that 1 gal = 0.1337 ft^3. Using the method described in Chapter 3 and rounding the answers, we find:

$$\frac{22{,}800\ \cancel{\text{ft}^3}}{1} \cdot \frac{1\ \text{gal}}{0.1337\ \cancel{\text{ft}^3}} = 171{,}000\ \text{gal}$$

This conversion could have been done just as easily using the conversion factor 1 ft^3 = 7.48 gal.

$$\frac{22{,}800\ \cancel{\text{ft}^3}}{1} \cdot \frac{7.48\ \text{gal}}{1\ \cancel{\text{ft}^3}} = 171{,}000\ \text{gal}$$

To convert to liters, we must look up the conversion from gallons to liters, or preferably from ft^3 to liters, since the original measurements were in feet. We find 1 ft^3 = 28.32 L.

$$\frac{22{,}800\ \cancel{\text{ft}^3}}{1} \cdot \frac{28.32\ \text{L}}{1\ \cancel{\text{ft}^3}} = 646{,}000\ \text{L}$$

So the tank has a volume of 22,800 ft^3 or 171,000 gal or 646,000 L.

Sphere: The volume of a sphere of radius, r, is given by the formula:

$$V = \frac{4}{3}\pi r^3$$

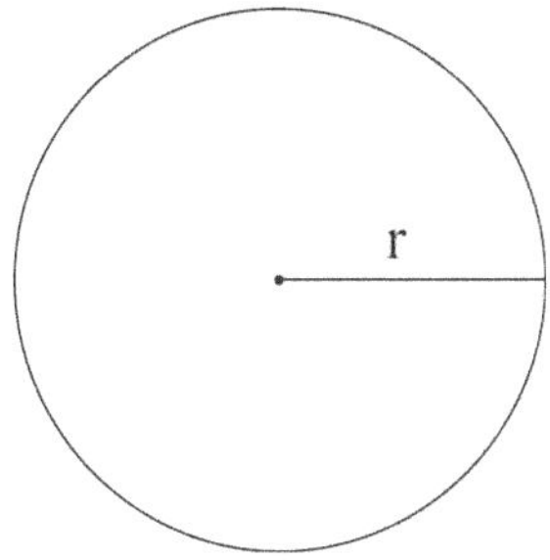

Notice that the radius is cubed (exponent of 3) so that the volume will have dimensions of ft^3, m^3, cm^3, etc. If we want the volume in liters, we must use a conversion factor.

The surface area of a sphere of radius, r, is given by the formula:

$$A = 4\pi r^2$$

Notice that the area varies as radius squared (exponent of 2) so that area will have dimensions of ft^2, m^2, cm^2, etc.

Example: What is the capacity, in liters, of a spherical tank with a diameter of 5.0 m? What is the surface area?
Solution: The volume and the area formulas are given in terms of radius, so we must find the radius. We know the diameter, which is twice the radius, so the radius is one half of the diameter, r = 2.5 m. Substituting into the volume formula and rounding the answer gives:

$$V = \frac{4}{3}\pi r^3$$

$$V = \frac{4}{3}\pi (2.5\text{ m})^3$$

$$V = 65\text{ m}^3$$

Converting to liters we get:

$$\frac{65\text{ m}^3}{1} \cdot \frac{1000\text{ L}}{1\text{ m}^3} = 65{,}000\text{ L}$$

So the tank has a capacity of 65,000 L or 65 kL.

Substituting into the area formula and rounding, we get:

$$A = 4\pi r^2$$

$$A = 4\pi (2.5 \text{ m})^2$$

$$A = 79 \text{ m}^2$$

The area of the spherical tank is 79 m^2.

4.3 Density

Fourth graders are often asked, "Which weighs more, a pound of feathers or a pound of lead?" The answer is that they both have the same weight, one pound. The difference is their density, or more strictly, their weight density. One pound of feathers occupies much more space than one pound of lead.

There are many kinds of density. Population density is usually expressed as so many people per square mile. A dermatologist may determine the number of hair follicles per square centimeter. The SSPC-VIS 2 rust grades are based on the amount of rust in a given area, or the surface density of rusting. The difference between the lead and the feathers is their weight density or mass density, which is usually what we mean when we say density. The density of paint is usually measured in lb/gal or g/L.

In this book, when we refer to density, we mean weight density or mass density. The difference between weight and mass is discussed in physics books. Briefly, weight is the force of gravity on the object, and mass is a measure of a body's inertia (how difficult it is to accelerate). Alternatively, we can think of mass as the amount of matter in an object. When we step on a scale, we measure our weight in pounds and our mass in kilograms. In outer space we will have zero weight but we will have the same mass as we have on earth. Weight depends on gravity, mass does not. Weight and mass are numerically related through the acceleration of free fall due to gravity, $g = 9.8\text{m/s}^2 = 32 \text{ ft/s}^2$. When we convert between pounds (weight) and kilograms (mass) we are really saying that a one kilogram mass has a weight of 2.205 pounds. Since the bulk of painting is done near the surface of the earth, the distinction between weight and mass becomes academic.

Weight (mass) density is defined as the ratio of weight (mass) to volume. We will use D to represent density. Often the Greek letter rho, ρ, is used for density. The formula for density can be rearranged algebraically to give:

Definition of Density

☞ $D = \frac{W}{V}$ OR $D = \frac{M}{V}$

where
W = weight
V = volume
M = mass

$$W = DV$$

and

$$V = \frac{W}{D}$$

The best approach is to remember only one formula, the definition, and use algebra to rearrange it as needed.

Example: If 5.0 gal of paint weighs 110 lb, what is its density?
Solution: We identify our known quantities as V = 5.0 gal and W = 110 lb. We are looking for density, D, so we substitute into the formula relating our knowns and our unknown.

$$D = \frac{W}{V}$$

$$D = \frac{110 \text{ lb}}{5.0 \text{ gal}}$$

$$D = 22 \frac{\text{lb}}{\text{gal}}$$

Example: The single component zinc-rich paint being used on the job has a density of 43 lb/gal according to the tech data sheet. The company pick-up truck has a load limit of one ton. How many gallons of paint can we legally carry in the truck? Ignore the weight of the cans.

Solution: We first identify the known quantities. The density, D, is 43 lb/gal. The total weight is one ton, which we must convert to 2000 lb so that our units are consistent. We are looking for volume (total number of gallons). Substituting into the density formula rearranged to find volume we get:

$$V = \frac{W}{D}$$

$$V = \frac{2000 \text{ lb}}{43 \, {}^{\text{lb}}\!/_{\text{gal}}}$$

$$V = 46.51 \text{ gal}$$

After rounding the answer we find that the legal payload is 47 gal.

To see how the units work out, divide lb by lb/gal. Remember that when we divide by a fraction, we "invert and multiply:"

$$\frac{\text{lb}}{\frac{\text{lb}}{\text{gal}}} = \frac{\frac{\text{lb}}{1}}{\frac{\text{lb}}{\text{gal}}} = \frac{\cancel{\text{lb}}}{1} \cdot \frac{\text{gal}}{\cancel{\text{lb}}} = \text{gal}$$

Specific Gravity: If we try to look up the density of some material in the *Handbook of Chemistry and Physics* or some other reference, we may not find a list of densities. Instead we will find a listing of specific gravity. Specific gravity is the ratio of the density of the material to the density of water.

$$\text{specific gravity} = \frac{\text{density of material}}{\text{density of water}}$$

$$\text{s.g.} = \frac{D}{D_{\text{water}}}$$

If the formula is rearranged we get:

$$D = (\text{s.g.}) \cdot D_{\text{water}}$$

Since specific gravity is a ratio, it is a dimensionless quantity, i.e., it has no units. Specific gravity is more fundamental than density because its value is the same in any system of units. The density of water is 62.4 lb/ ft^3 or 1000 kg/m^3 or 1 g/cm^3 or 1 g/mL or 1 g/cc. (Recall that 1 mL = 1 cm^3 = 1 cc and that 1 mL of water weighs 1 g.) The density in g/mL is numerically equal to the specific gravity. The specific gravity of mercury is 13.6, so one mL of mercury will weigh 13.6 g.

From the definition of specific gravity, the specific gravity of water is 1. If the specific gravity of a substance is 2, that means that it will weigh twice as much as an equal volume of water. Mercury is 13.6 times as dense as water. A barometer filled with water would have to be 13.6 times taller than one filled with mercury.

Example: A handbook gives the specific gravity of steel as 7.83. What is the density of steel?
Solution: Substituting into the rearranged definition of specific gravity:

$$D = (\text{s.g.}) \cdot D_{\text{water}}$$

$$D = (7.83) \cdot 62.4 \frac{\text{lb}}{\text{ft}^3}$$

$$D = 488.592 \frac{\text{lb}}{\text{ft}^3}$$

So the density of steel (after rounding) is 489 lb/ft^3.

The density in g/cm^3 is numerically equal to the specific gravity, or 7.83 g/cm^3. Using the density of water as 1000 kg/m^3, we move the decimal point three places to find the density of steel to be7830 kg/m^3.

Example: What is the weight of the steel H-beam described in the earlier example if the density of steel is 489 lb/ft^3? (See page 54.)
Solution: We know the density and we can find the volume. Then we can use the formula W = D·V to find the weight. The volume of the beam is the cross-sectional area times the length. We convert the area of 19 in^2 into ft^2 so our units are consistent.

$$\frac{19\ \cancel{\text{in}^2}}{1} \cdot \frac{1\ \text{ft}^2}{144\ \cancel{\text{in}^2}} = 0.132\ \text{ft}^2$$

We multiply this by the length to get the volume of the beam.

$$V = A \cdot L$$
$$V = (0.132\ ft^2)(20\ ft)$$
$$V = 2.64\ ft^3$$

Continue substituting

$$W = D \cdot V$$
$$W = (489 lb/\cancel{ft^3})(2.64\ \cancel{ft^3})$$
$$W = 1290 lb$$

So the weight of the beam is 1290 lb. If the first digit of a number is 1, it is permissible to carry an extra significant figure. Significant figures are discussed in depth in the next chapter.

Example: A tank with an area of 24,000 square feet was cleaned with ultra-high pressure waterjetting (UHP WJ). The average coating thickness was 18 mils. Estimate the weight of the dried paint debris. Dry paint usually has a specific gravity between 1.2 and 2.5. Use the value of 2.0 in this computation.
Solution: We can find the total weight if we know the volume of the paint and the weight of a unit volume (i.e., the density).

We first find the volume of paint. We know the area and the thickness. The general formula gives volume as the area of the base times the height. In this problem, the height is simply the paint thickness, 18 mils. In order to get the volume in cubic feet, we must convert mils to feet.

$$\text{Volume} = (\text{area of base}) \cdot (\text{height})$$
$$V = (24{,}000\ ft^2) \cdot (18\ \cancel{mils}) \cdot \frac{1\ \cancel{in}}{1000\ \cancel{mils}} \cdot \frac{1\ ft}{12\ \cancel{in}}$$
$$V = 36\ ft^3$$

Thus, there was about 36 ft^3 of paint removed from the tank.

We can compute the density of dry paint in lb/ft³ from the definition of specific gravity and the fact that the density of water is 62.4 lb/ft³.

$$D = (\text{s.g.}) \cdot D_{\text{water}}$$

$$D = (2.0) \cdot \left(62.4 \frac{\text{lb}}{\text{ft}^3}\right)$$

$$D = 125 \frac{\text{lb}}{\text{ft}^3}$$

We use the volume and the density to find the total weight,

$$W = V \cdot D$$

$$W = (36\,\cancel{\text{ft}^3}) \cdot \left(125 \frac{\text{lb}}{\cancel{\text{ft}^3}}\right)$$

$$W = 4500 \text{ lb}$$

Thus, the weight of paint debris from this job will be about 4500 lb.

This computation assumed that the paint waste was dry (no water) and that 100% of the paint was collected. The pile of waste will occupy a volume of more than 36 ft³ because of the air space between the paint chips. We could always weigh a gallon of waste and determine the bulk density of the waste in pounds per gallon. From this bulk density and the size of a disposal drum, we could determine the number of drums we would need.

Chapter 5 – Computations

Objective: To learn how many digits we should keep in our answer when the display on a calculator consists of 8 or more digits; how to write very large and very small numbers without using a lot of zeros; and how to find and use percentages, among other challenges related to computations.

5.1 Rounding

Suppose we put data for a paint job into a formula and the calculator gives 65,824 pounds of abrasive needed. We would estimate that we would need about 66,000 pounds, or 33 tons (since 1 ton = 2000 lb). We have rounded 65,824 to the nearest thousand as 66,000, because 65,824 is closer to 66,000 then to 65,000.

The rules for rounding are on page 69. A few examples listed in Table 5.1 should make these rules more understandable.

5.2 Significant Figures

The supervisor estimates to the nearest hundred that it will take 500 bags of coal slag abrasive to blast clean a ramp. This means that the actual amount needed will be closer to 500 than to 400 or 600. In other words, the number of bags needed will be between 450 and 550. If we needed less than 450, the estimate would have been 400 bags; and if we needed more than 550, the estimate would have been 600 bags. Merely stating the number as 500 means between 450 and 550. Only the 5 in 500 is significant. The two zeros are place holders to tell us that the 5 is in hundreds. We say that 500, used in this sense, has one significant figure.

By looking at a map, we determine that a job is about 200 miles from the shop. From our estimate we believe that the job is between 195 and 205 miles away. If it was less than 195 miles, we would have said the distance was 190 miles; and if it was more than 205 miles, we would have estimated 210 miles. Our estimate is good to the nearest 10 miles, i.e., we are counting by tens. The last zero is not significant, because if we measured the distance more precisely we may find it to be 203 miles or 198 miles; both of which can be rounded to 200 miles. In this example, 200 has two significant figures, the 2 and the first zero (in the tens place). The significance of the first zero is sometimes indicated by putting a bar over

Table 5.1

Examples of Rounding

EXACT	ROUNDED TO THE NEAREST		
	TEN	HUNDRED	THOUSAND
47,295	47,300	47,300	47,000
125,251	125,250	125,300	125,000
4,525	4,520	4,500	5,000
36,015	36,020	36,000	36,000
7,165	7,160	7,200	7,000
423.91	420	400	0

EXACT	ROUNDED TO THE NEAREST		
	UNIT	TENTH	HUNDREDTH
6.512	7	6.5	6.51
86.926	87	86.9	86.93
0.055	0	0.1	0.06
29.965	30	30.0	29.96
0.7924	1	0.8	0.79

the significant zero, i.e., $2\bar{0}0$.

Our records show that there were 300 hours spent on the job last week. In this case, 300 means between 299.5 and 300.5. If the time was less than 299.5, we would have recorded the time as 299; and if it was more than 300.5, we would have recorded 301. The number, 300, is rounded to the nearest hour. The last zero means zero and cannot be 1 or 2 or 9. In this instance, 300 has three significant figures, the 3 and both zeros.

If the tech data sheet says that the VOC concentration is 3.6 g/L, it means that it is between 3.55 and 3.65 g/L. If it was less than 3.55, the stated value would be 3.5 g/L; and if it was more than 3.65, the stated value would be 3.7 g/L. The number 3.6 has two significant figures.

If the VOC concentration is listed as 3.60 g/L, then the value must be between 3.595 and 3.605. The number 3.60 has three significant figures.

If the shim thickness is marked as 0.040 inches, only the 4 and the last zero are significant. The beginning zeros are needed because the measurement was in inches. This thickness could also be expressed as 40 mils, since 1 mil = 1/1000 inch. The shim thickness is closer to 40 mils than it is to 39 mils or 41 mils. That is why we need the last zero. The number, 0.040, has two significant figures.

If we were to write 0.04 inches, we would assume the measurement is between 0.035 and 0.045 inches. The 0.04 inches can be converted to 40 mils. In this case it means that the measurement is closer to 40 mils than it is to 30 mils or 50 mils. The number, 0.04, has only one significant figure.

The examples just stated should help us to get a feel for the rules determining the number of significant figures. More examples are given in Table 5.2.

In the examples at the beginning of this section, how should we write 200 miles to show two significant figures and 300 work-hours to

Rules for Rounding

- If the next digit is less than 5, change that digit and those following to zero.
- If the next digit is 5 (followed by at least one non-zero digit) or greater, increase the digit to which we are rounding by one and change the following digits to zero.
- If the next digit is 5 followed by all zeros (if there are any more digits), leave the previous digit as is or add one to the previous digit so that the last digit of the rounded number is even.
- If the next digit is after the decimal point, it and the following digits are dropped.

Table 5.2

Number of Significant Figures

NUMBER	NUMBER OF SIGNIFICANT FIGURES
65.7	3
4500	2
5007	4
3.50	3
0.009	1
50.0	3
2.000	4

show three significant figures. According to the rules, 200 and 300 have only one significant figure. We can know the level of significance from the context in which they are used, we can include the decimal point (300.), we can use a bar over the significant zero $3\bar{0}\bar{0}$, or we can use scientific notation (2.0×10^2 or 3.00×10^2) [see Section 5.3]. Scientific notation is the preferred method.

Significant Figures in Addition and Subtraction: To add or subtract decimals by hand, we line up the decimal points and then add or subtract the numbers.

If desired, we can add zeros at the end of the numbers. After performing the operations, the answer should be rounded to the same number of places as the least precise number in the list. The rules for rounding the final answer are the same whether or not a calculator is used.

Example: Add 3.65 + 24.7 + 0.892.

Solution: List the numbers with the decimal points lined up and perform the computation.

3.65	3.650
24.7	24.700
0.892	0.892
29.242	29.242

The 24.7 is only precise to the nearest tenth, so the sum (the answer in addition) can only be precise to the nearest tenth. Hence, round the final answer to 29.2. It is good practice to do the computation with all the significant figures and then round the final answer.

Rules for Significant Figures

- All non-zero digits are significant
- Zeros between significant figures are significant
- All zeros at the beginning of a number are not significant
- Zeros at the end of a whole number are not significant
- Zeros at the end of a number after the decimal point are significant

Example: Subtract 5,800 from 46,320.
Solution: List the numbers with the decimal points lined up and perform the computation. The decimal points at the end of a whole number are not usually written.

$$
\begin{array}{r}
46{,}320 \\
-5{,}800 \\
\hline
40{,}520
\end{array}
$$

Since the number, 5800, is only precise to the nearest hundred (it has only two significant figures), we round the answer to the nearest hundred and state it as 40,500.

- When adding and subtracting, round the answer to the (decimal) place precision of the least precise number in the list.

Significant Figures in Multiplication and Division: The volume of a box is computed by multiplying the length times the width times the height, $V = L \cdot W \cdot H$.

Suppose L = 35.2 cm, W = 15.71 cm, and H = 6.736 cm. Multiplying these numbers on a calculator, we find the volume is 3,724.9541 cm^3. Clearly, we do not know the volume this precisely. How many significant figures should we keep? Where do we round? The rule developed by mathematicians is:

☞ When multiplying and dividing, keep as many significant figures in the answer as there are in the least precise factor.

In this example, the width and the height are known to four significant figures, but there are only three significant figures in the length; so the volume should be written with only three significant figures, V = 3,720 cm^3.

The number of significant figures used in recording a measurement indicates the level of sensitivity of the measurement, which is usually limited by the measuring instrument. Another way to indicate uncertainty in a measurement is to explicitly state the range of uncertainty. For example, if the length of a beam is measured to the nearest half-foot, the length could be recorded as 34 ± 0.5 ft, which means that the actual length of the beam is between 33.5 ft and 34.5 ft. If the measurement is only to the nearest foot, the length could be recorded as 34 ± 1 ft, which indicates an actual length between 33 ft and 35 ft.

5.3 Scientific Notation

Scientific notation is used to write very large or very small numbers, and to eliminate ambiguities as to the number of significant figures. A number such as 5 billion can be written with nine zeros 5,000,000,000 or in scientific notation as 5×10^9. The exponent, 9, tells us to multiply by 10 nine times. Each time we multiply by 10 we move the decimal point one place to the right. Thus, if we multiply by 10 nine times, we move the decimal point nine places to the right. In this case, each time we move the decimal point, we must add a zero.

To write a number in scientific notation, we write the first non-zero digit, then the decimal point, then the rest of the significant figures, then multiply by 10 to the appropriate power. A number in scientific notation consists of a number between 1 and 10 multiplied by a power of 10.

We can write very small numbers, such as 0.000 006, in scientific notation if we use negative exponents, 6×10^{-6}. The negative exponent means that we must divide by 10 six times. Each time we divide by 10, we move the decimal point one place to the left. In this case we move the decimal point six places to the left. Scientific notation is useful when converting units in the metric system.

Example: Convert 0.000 006 meters to micrometers.
Solution: We look up the metric prefix and find that micro means 10^{-6}, so $1\ \mu m = 10^{-6}$ m.

$$0.000006\ m = \frac{6 \times 10^{-6}\cancel{m}}{1} \cdot \frac{1\mu m}{10^{-6}\cancel{m}} = 6\mu m$$

So 0.000 006 meters is equal to 6 micrometers.

Scientific notation is not usually used unless the numbers are very small or very large. Table 5.3 may help us see the relationship between scientific notation and ordinary notation.

Table 5.3

Comparison of Decimal Notation with Scientific Notation

DECIMAL NOTATION	SCIENTIFIC NOTATION	NUMBER OF SIGNIFICANT FIGURES
2500	2.5×10^{3}	2
2500	2.50×10^{3}	3
2500	2.500×10^{3}	4
0.038	3.8×10^{-2}	2
0.0380	3.80×10^{-2}	3
49.0	4.90×10^{1}	3
485,000	4.85×10^{5}	3

Returning to our earlier examples, if 200 miles has two significant figures, it can be written in scientific notation as 2.0×10^2 mi. If 300 worker-hours has three significant figures, it can be written as 3.00×10^2 worker-hours. In these two examples, scientific notation is more cumbersome, but it is also less ambiguous.

Computations with numbers written in scientific notation can be done on a calculator in the same fashion as regular numbers. We only have to learn how to enter numbers in scientific notation and how to read the output.

To enter 5.3×10^6, follow these steps.

☞ Enter 5.3.

☞ Press the "EE" or "EXP" button. At this time 00 should appear on the right end of the display separated from the 5.3.

5.3	00

☞ Enter the exponent, 6. The display should now have 5.3 followed by some space and an "06."

5.3	06

To enter 4.9×10^{-5}, follow these steps.

☞ Enter 4.9.

☞ Press the "EE" or "EXP" button.

☞ Enter 5.

☞ Press the +/– button. This makes the 5 negative. The display should now show 4.9 followed by some space and "–05."

4.9	–05

The EE or EXP button automatically adds the base 10 as a multiplier, and all we have to do is enter the exponent. There is no need to enter the multiply button, X, or enter a 10 followed by the x^y or y^x button.

Example: Divide (5.3 x 10^6) ÷ (4.9 x 10^{-5})
Solution: Follow these steps.

ENTER	DISPLAY	COMMENT
5.3	5.3	
EE or EXP	5.3 00	puts number in scientific notation
6	5.3 06	makes 6 the exponent on 10
÷	5.3 06	divide
4.9	4.9	
EE or EXP	4.9 00	puts number in scientific notation
5	4.9 05	makes 5 the exponent on 10
+/–	4.9 –05	makes the exponent negative
=	1.0816 11	the answer

The answer can be rounded to 1.08 x 10^{11}.

5.4 Percents

Percent means "by one hundred" or "per hundred". If 35 feet of a hundred foot railing needs to be painted, then 35 percent (35%) of the railing needs to be painted. We are probably familiar with the meaning of some more common percentages. For example, a 50% off sale means half price, 25% means one fourth, 33 1/3% means one third, and 90% means 9 out of 10. Others are not as easy to interpret, such as, 87.4% solids. In this section we will see how percentage is simply another way of writing a number. We will show how to convert between fraction, decimal, and percent notation. Then we will work some problems involving percentages.

Since percent means per hundred, any percent can be expressed as a fraction with the denominator being 100. For example, 35% means 35/100, 4.8% means 4.8/100, 0.6% means 0.6/100, 14 2/7% means 14 2/7 over 100, 250% means 250/100 and so on. Let's look at each of these percentages and write them in different forms.

• 35% by definition is 35/100. If we now reduce this fraction to lowest terms by dividing the numerator and the denominator by 5 we get 7/20. Thus, 35% = 7/20. Remember that 35/100 means 35 divided by 100, and to divide by 100 we move the decimal point two places to the left getting 0.35. Thus, 35% = 0.35. Therefore, 35% = 35/100 = 7/20 = 0.35.

• Using similar techniques we can convert 4.8% to other forms.

$$4.8\% = \frac{4.8}{100} \cdot \frac{10}{10} = \frac{48}{1000}$$

Reducing this fraction to lowest terms by dividing the numerator and the denominator by 8 gives 6/125. Moving the decimal point to the left we get 0.048 as the decimal equivalent. Thus, 4.8% = 6/125 = 0.048.

• Continuing in this fashion, 0.6% can be converted to a fraction, 3/500, or a decimal, 0.006.

$$0.6\% = \frac{0.6}{100} \cdot \frac{10}{10} = \frac{6}{1000} = \frac{3}{500} = 0.006$$

• In converting 14 2/7% we must express 14 2/7 as an improper fraction. To get the numerator of the improper fraction we multiply 14 by 7 and add 2, which is 100. The denominator remains 7. Thus, 14 2/7% is 100/7%. Putting this over 100 results in a complex fraction because the numerator is itself a fraction.

$$14\frac{2}{7}\% = \frac{100}{7}\% = \frac{\frac{100}{7}}{100} = \frac{\cancel{100}}{7} \cdot \frac{1}{\cancel{100}} = \frac{1}{7}$$

Because 1/7 is not a terminating decimal (it is a repeating decimal) we will find a decimal approximation. Dividing 1 by 7 on a calculator we find 1/7 = 0.142857. This sequence of digits keeps repeating. Rounding to the third decimal place we get 14 2/7% ≈ 0.143. (The symbol ≈ means "approximately equal to.") An alternate approach is to convert 2/7 into a decimal 2/7 = 0.2857143 ≈ 0.29, so that 14 2/7% ≈ 14.29%. We then may proceed as in the previous examples.

$$14\frac{2}{7}\% \approx 14.29\% = \frac{14.29}{100} = 0.1429 \approx 0.143$$

• Converting a number more than 100%, such as 250%, sometimes creates problems. Using the same methods as we used above we can express 250% as an improper fraction, a mixed number, or a decimal.

$$250\% = \frac{250}{100} = \frac{5}{2} = 2\frac{1}{2} = 2.5$$

Now that we can go from a percent to a fraction or a decimal, we start with a fraction or mixed number and transform it to a decimal or a percent. To convert from a fraction to a decimal, we simply divide the numerator by the denominator. Dividing 5 by 8 we get 5/8 = 0.625. If the decimal does not terminate, as this one did, we round to the desired number of places. Once we have the decimal equivalent, we need only move the decimal point two places to the right to get the percent, 0.625 = 62.5%.

Another method to use in converting among percents, decimals, and fractions uses the fact that 100% = 100/100 =1. We now try to understand why moving the decimal point two places to the right will convert a decimal to a percent, and how to determine that 1/3 is exactly 33 1/3%. We can always multiply or divide by one without changing the value of the quantity. So if we start with a fraction, such as 4/5, and multiply by one (100%) we get:

$$\frac{4}{5}\cdot 1 = \frac{4}{5}\cdot 100\% = \left(\frac{4}{\cancel{5}_{1}}\cdot\frac{\overset{20}{\cancel{100}}}{1}\right)\% = 80\%$$

We can always convert 4/5 to a decimal before finding the percent.

$$\frac{4}{5} = 4 \div 5 = 0.8$$

$$0.8 \cdot 1 = 0.8 \cdot 100\% = (0.8 \cdot 100)\% = 80\%$$

Notice that in multiplying 0.8 by 100 we merely had to move the decimal point two places to the right.

Example: Express 5/7 as an exact percent.

Solution: We multiply 5/7 by one in the form of 100%.

$$\frac{5}{7}\cdot 1 = \frac{5}{7}\cdot 100\% = \left(\frac{5}{7}\cdot\frac{100}{1}\right)\% = \frac{500}{7}\% = 71\frac{3}{7}\%$$

So 5/7 = 71 3/7%.

To get a decimal approximation, convert 5/7 to a decimal and then move the decimal point two places to the right.

$$\frac{5}{7} \approx 0.7143 = 0.7143 \cdot 100\% = (0.7143 \cdot 100)\% = 71.43\%$$

So 5/7 ≈ 71.43%

If we ever get confused on whether to move the decimal point two places to the right or two places to the left, we can either multiply or divide by 100% as needed, since we are multiplying or dividing by one.

Example: Convert 0.02% to a decimal.
Solution: Since we know the percent, we are trying to eliminate it. Therefore we want to divide by 100%.

$$0.02\% = \frac{0.02\%}{1} = \frac{0.02\%}{100\%} = \frac{0.02}{100} = 0.0002$$

The % cancels; and when we divide by 100, we must move the decimal point two places to the left. So, 0.02% = 0.0002.

Since numbers can be expressed as a decimal, a percent, or a fraction, it is useful to be able to convert from one form to any other form. To convert a decimal to a fraction, write the decimal with a denominator of one, then multiply the numerator and the denominator by 10 as often as necessary to eliminate the decimal. Finally, we reduce the fraction to lowest terms.

Example: Convert 0.55 to a fraction and to a percent.
Solution: Write 0.55 as a fraction with a denominator of 1 and multiply numerator and denominator by 100.

This is equivalent to moving both decimal points two places to the right. Then reduce the fraction to lowest terms.

$$0.55 = \frac{0.55}{1} \cdot \frac{100}{100} = \frac{55}{100} = \frac{11}{20}$$

Notice that 0.55 and 55/100 are both read "fifty-five hundredths."

To convert to a percent, multiple 0.55 by 100%.

$$0.55 \text{ x } 100\% = 55\%$$

So, 0.55 = 11/20 or 55%. With a little practice, many of these steps can be done mentally. They have been written out here in order to illustrate the thought process.

Rule for Changing to and from a Percent

Since 100% = 1,

- ☞ To change TO a percent, multiply by 100%
- ☞ To change FROM a percent, divide by 100%

Summary of steps for converting percents:

- To convert from a fraction to a percent, multiply the fraction by 100 and add the percent sign.
- To convert from a decimal to a percent, move the decimal point two places to the right and add the % sign.
- To convert from percent to decimal, move decimal point two places to the left and eliminate % sign.
- To convert from percent to fraction, divide by 100%, cancel the % signs, reduce the fraction to lowest terms.

There are many different ways of attacking problems involving percents. Most people prefer the first method they understood. The method presented here is closely related to the material presented in the other chapters and is consistent with the method for solving any word problem. Simple problems involving percents can be paraphrased into the form:

_____ is _____% of _____ OR _____% of _____ is _____

These two forms are mirror images of each other. In any problem we must know numbers for two of the three blanks, and then use algebra to solve for the piece we do not know. The method consists of translating the statement into an equation and solving for the unknown. A few examples will illustrate the method.

Example: What is 43% of 81?
Solution: The first step is to recognize the type of problem as a percentage problem. We know two of the three pieces, 43% and 81. We do not know the "what." The simplest way to solve problems is to assign a letter to the piece we do not know. Let's let W represent the "what." When translating from a word statement (a sentence) into a mathematical equation "is" means "equal to" and "of" means "times" or multiply.

What is 43% of 81?

$$W = 0.43 \cdot 81$$

The percent must be converted to a decimal or a fraction before substituting into the equation. The equation can be written:

$$W = (0.43) \cdot (81)$$

Multiplying on the calculator and rounding to two significant figures gives W = 35. So 43% of 81 is 35.

Example: 24 is 150% of what number?
Solution: We assign a letter, N, to the quantity that we do not know. It can be less confusing to use a letter representative of the quantity in question, in this case N for number, rather than always using x as the unknown. Any letter will do perfectly fine. We must convert 150% to 1.50. Next we translate the sentence into an equation.

24 is 150% of what number?

$$24 = 1.50 \cdot N$$

Writing the equation normally we get:

$$24 = 1.50\text{ N} \quad \text{OR} \quad 1.50\text{ N} = 24$$

Dividing both sides by 1.50 we get:

$$\frac{1.50\text{ N}}{1.50} = \frac{24}{1.50}$$

$$\text{N} = 16$$

So 24 is 150% of 16.

We knew the percent in the previous two examples. In this next example we do not know the percent.

Example: What percent of 62 is 18?
Solution: We do not know the percent, so we assign a letter, P, to represent the percent. Next we translate into an equation.

What percent of 62 is 18?

$$\text{P} \cdot 62 = 18$$

Writing in a condensed form we get:

$$\text{P}\cdot 62 = 18 \quad \text{OR} \quad 62\text{ P} = 18$$

We divide both sides by 62 to find P (in decimal form).

$$\frac{\cancel{62} \cdot \text{P}}{\cancel{62}} = \frac{18}{62}$$

$$\text{P} = 0.2903226 \approx 0.29$$

We now convert from a decimal to a percent:

$$\text{P} = 0.29 \cdot 100\% = 29\%$$

So 18 is 29% of 62.

More Percent Problems

Example: A vendor at the SSPC Exhibition is offering a 15% discount on a

DFT gauge that lists for \$789. How much will we save by buying the gauge at the conference? What price will we pay for the gauge?

Solution: This problem deals with percents. The first step is to paraphrase the question so that it is in the form of the previous examples, i.e. a sentence with the verb "is."

Let us call the amount saved, the discount. We can then restate the problem: The discount is 15% of \$789. Using the letter D to represent the discount, we translate the sentence into an equation, $D = 0.15 \cdot 798$. (The 15% was written as 0.15 and is an exact percent, which means 15.000…%.) Doing the computation on a calculator and rounding to the nearest cent gives $D = 118.35$. Thus, our savings on the gauge will be \$118.35.

Our cost for the gauge will be the list price less the discount. This part may be done mentally if we are familiar with this type of problem, but we will do it algebraically to illustrate the algebraic process. We paraphrase with an "is" type statement, assign letters to the quantities, write the equation, substitute for the things we know, and solve for the unknown. If our cost is C, the discount is D, and the list price is L, we have:

Cost is list price less discount.

$$C = L - D$$

$$C = 789 - 118.35$$
$$C = 670.65$$

Our cost for the gauge with the 15% discount will be \$670.65.

Alternative solution: There are often several correct approaches to solving mathematical problems. We can reason that if we paid full price for the gauge, we would be paying 100% of the list price. Since the discount is 15%, we can subtract this from 100% to get 85%. Our cost will be 85% of the list price. Writing an "is" type statement, translating, and substituting we get:

Cost is 85% of \$789.

$$C = 0.85 \cdot 789$$

$$C = 670.65$$

The same as we got above.

To find the discount in dollars we form the statement:

Discount is list less cost.

$$D = L - C$$
$$D = 789 - 670.65$$

$$D = 118.35$$

Although we do not have to write all of these steps down, this is the reasoning path our mind will probably take to solve the problem.

Example: A new GMC Sierra pickup, equipped as we like it, lists for \$28,564. The dealer will sell it to us for \$22,500. What percent of the list price was our discount?
Solution: Writing an "is" statement we may say, "The discount is what percent of the list?" Using letters for discount (D), percent (P), and list (L) this becomes $D = P \cdot L$. The only one of these letters we know is $L = 28{,}564$. Before solving this equation we must know either D or P, because we must know all the letters but one. We can find the discount by subtracting the cost (C) from the list price.

$$D = L - C$$
$$D = 28{,}564 - 22{,}500$$
$$D = 6064$$

Now that we know D we can substitute this into our equation:

$$D = P \cdot L$$
$$6064 = P \cdot 28{,}564$$

Dividing both sides of the equation by 28,564 we get:

$$\frac{6064}{28{,}564} = \frac{P \cdot 28{,}564}{28{,}564}$$
$$0.212295 = P$$

OR

$$P \approx 0.21$$

Converting this to a percent:

$$P = 0.21 \cdot 100\% = 21\%$$

Thus, the dealer is selling us the truck at a 21% discount.

Alternative solution: Another approach is to determine the percent of the list price we paid, then subtract this from 100%. Forming an "is" statement and substituting we find:

$$\begin{array}{lccc} 22{,}500 \text{ is what percent of } 28{,}564? \\ 22{,}500 = & P & \cdot & 28{,}564 \end{array}$$

Dividing both sides by 28,564 we get:

$$\frac{22{,}500}{28{,}564} = \frac{P \cdot 28{,}564}{28{,}564}$$

$$0.7877 = P$$

$$\text{OR}$$

$$P \approx 0.79$$

Converting to a percent:

$$P = 0.79 \cdot 100\% = 79\%$$

So our cost was 79% of the list. Subtracting 79% from 100% gives us a discount of 21%.

Example: The manufacturer of SuperGrit claims his product cleans 20% faster than silica sand. If it took 16 hours to clean the structure with SuperGrit, how long would it have taken with sand?
Solution: Cleaning 20% faster than sand means that it took 80% as long to clean with SuperGrit (100% – 20% = 80%). Since we do not know how long it would take with sand, we let T equal the time to clean with sand. We write an "is" statement, "16 hours is 80% of the time to clean with sand." Writing this as an equation:

$$16 = 0.80 \cdot T \qquad \text{(80\% was written as 0.80)}$$

Dividing both sides by 0.80 we get:

$$\frac{16}{0.80} = \frac{0.80 \cdot T}{0.80}$$

$$20 = T$$

So it would take 20 hours to clean with sand.

Chapter 6 – Statistics

Objective: To review elementary statistical quantities.

6.1 Central Tendency

Statistics is a method of describing a set of numbers or measurements with only a few numbers. The three most common statistics used to describe the center of a set of measurements are the mean, the median, and the mode. These statistics, each computed in a different manner, consist of a single value (except perhaps the mode) that best represents the data. The mode is usually a single value, but it may be more than one value or may not exist at all. In many cases, all three statistics are used to give a complete description of the data.

The **mean**, more exactly the arithmetic mean, is what we normally refer to as the average value. If Bob weighs 200 lb and Joe weighs 150 lb, their average weight is 175 lb. The average is obtained by adding the weights (200 + 150 = 350) and then dividing by the number of people; so the average is 350 ÷ 2 = 175 lb.

If the five spot measurements (in mils) for DFT are 2.4, 2.7, 1.9, 2.2, and 2.5, we find the mean DFT by computing the average of these five measurements. The average is (2.4 + 2.7 + 1.9 + 2.2 + 2.5) ÷ 5 = 11.7 ÷ 5 = 2.34. The average DFT may be rounded to the nearest tenth and given as 2.3 mils.

Although most scientists express the average to the same precision as the data, statisticians keep an extra significant figure when computing averages. If we were to use statistical functions to compare this DFT with another DFT determination, the DFT in our example above would be given as 2.34 mils. (See Section 6.2 for computing standard deviation and confidence intervals.) However, we are not sure that the DFT is between 2.335 and 2.345, which is the implication of writing 2.34 mils with two decimal places. We are more confident that the true DFT is between 2.25 and 2.35 mils, so we express DFT using only one decimal place as 2.3 mils.

The **median** is the number from the middle of a list that is arranged in ascending (or descending) order. There are as many numbers higher than the median as there are lower than the median. In the list of DFTs given above, the median would be 2.4, because there are two numbers greater than 2.4 (2.5 and 2.7) and two numbers less than 2.4 (2.2 and 1.9). If there are an even number of data points, the median is the mean (average) of the two middle data points. In this case, the median may be a number not corresponding to an actual data point.

The **mode** is the measurement that occurs more than any other measurement. In this set of data there is no mode because no single measurement occurs more frequently; each measurement occurs only once. The mode is meaningful for larger sets of numbers. A set of numbers may be bimodal, which means that there are two modes, or multimodal. For example, if we measured the height of everyone in a college auditorium, we may find that 5'9" and 5'4" were the most common heights. This could be because 5'9" is the most common height for men while 5'4" is the most common for women.

Example: The DFT of each panel in a set of 20 replicate test panels is given in Table 6.1. Find the mean, the median, and the mode of these measurements.

Solution: The first step is to list the data in order. If the data are recorded on a spreadsheet, the computer can sort the data easily. Select the column of data that is to be sorted and click the sort button in the tool bar. The sort button has an A over a Z with an arrow beside it pointing downward. Data can also be sorted by selecting "Data" in the tool bar and then choosing "Sort" from the Data menu.

The mean (average) is obtained by adding up the DFTs and dividing by 20. The sum of the DFTs is 68.4, so the average DFT is 68.4 ÷ 20 = 3.42 (which can be rounded to 3.4). We will use the symbol x_{avg} to represent the average value; $x_{avg} = 3.42$. Other notations for average value are x' and $\bar{x}$ (pronounced x bar).

ASTM's *MNL7A Presentation of Data and Control Chart Analysis* stipulates that when computing averages it is permissible to use an additional significant figure. It is good practice to carry extra significant figures when doing intermediate computations. Thus, we should use $x_{avg} = 3.42$ in the formula for standard deviation described in the next section.

The median is the measurement of the panel halfway down the sorted list. Since there is an even number of panels, numbers 10 and 11 are both equidistant from the ends of the list. The median is the average of these two DFTs. These panels both have DFTs of 3.5, so the median is 3.5 mils. Microsoft Excel or other spreadsheet software can be used to find the median without sorting the data. The mode is the most frequent measurement. In this list it is 3.5 mils (5 times).

Table 6.1

RAW DATA		SORTED	
PANEL #	DFT	PANEL #	DFT
1	3.1	7	2.9
2	3.4	14	3.0
3	3.5	1	3.1
4	3.5	8	3.1
5	3.5	15	3.2
6	3.4	2	3.4
7	2.9	6	3.4
8	3.1	9	3.4
9	3.4	13	3.4
10	3.7	3	3.5
11	3.5	4	3.5
12	3.6	5	3.5
13	3.4	11	3.5
14	3.0	20	3.5
15	3.2	12	3.6
16	3.7	18	3.6
17	3.8	19	3.6
18	3.6	10	3.7
19	3.6	16	3.7
20	3.5	17	3.8
		SUM	68.4

6.2 Dispersion

The spreading out or bunching together of the data is called dispersion. A 20 lb baby and a 200 lb father have an average weight of 110 lb. Two women, 100 lb and 120 lb, also have an average weight of 110 lb. So average value, alone, does not tell us how the data are distributed. The most common measures of dispersion are the range and the standard deviation.

The **range** is the difference between the largest data point and the smallest data point. In the baby/father example, the range is 180 lb; with the women the range is 20 lb. With small numbers of data points, range is a useful measure of dispersion.

In the test panel DFT data, the range is 3.8 – 2.9 = 0.9 mils. If 40 panels had been prepared instead of 20 panels, the range would most likely increase, since there would be a greater probability of having a panel with a DFT higher than 3.8 mils or lower than 2.9 mils.

The most common statistic for measuring dispersion is the **standard deviation**. Here we will only talk about the sample standard deviation. The standard deviation, s, of the 20 DFT readings in the example above is 0.24. We will discuss the meaning of this number and then later in this section show how to compute it. The average DFT was $x_{avg} = 3.42$ mils and $s = 0.24$. If another panel were to be prepared, there would be about a 68% probability that its DFT would lie within one standard deviation of the mean. The mean less one standard deviation is $x_{avg} - s = 3.42 - 0.24 = 3.18$; the mean plus one standard deviation is $x_{avg} + s = 3.42 + 0.24 = 3.66$. There is a 68% probability that the DFT of the next panel would be between 3.18 and 3.66 mils.

If we broaden the limits over which we may expect the value of the next measurement to fall and go two standard deviations from the mean, the probability of falling in this range increases to about 95%. Computing this range we get $x_{avg} + 2s = 3.42 + 2(0.24) = 3.90$ as the upper limit and $x_{avg} - 2s = 3.42 - 2(0.24) = 2.94$ as the lower limit. Thus, there is a 95% probability that the DFT of the next panel will be between 2.94 and 3.90 mils.

In a similar fashion, there is about a 99.7% probability that the next reading will fall within three standard deviations of the mean, i.e., between 2.70 and 4.14 mils. If a measurement is more than three standard deviations from the mean, it is often rejected as an "outlier." Using the list of DFT data, if we were to measure the DFT of a spot to be 4.5 mils, there is a high probability that the probe tip was sitting on a spec of dust, the measurement technique was faulty, or some other anomaly was present. Caution must be exercised in rejecting seemingly high (or low) data points. If the high (or low) measurement cannot be repeated, there is cause for rejection.

The probabilities 68%, 95%, and 99.7% are based on the assumption that the data are approximately normally distributed. In a normal distribution, the mean, the median, and the mode are the same number. This means that an individual data point is just as likely to be greater than the mean as it is to be lower than the mean. DFT measurements can be considered an essentially normal distribution. A normal distribution is often referred to as bell-shaped. Figures 1 and 2 show the effect of standard deviation on data that are normally distributed.

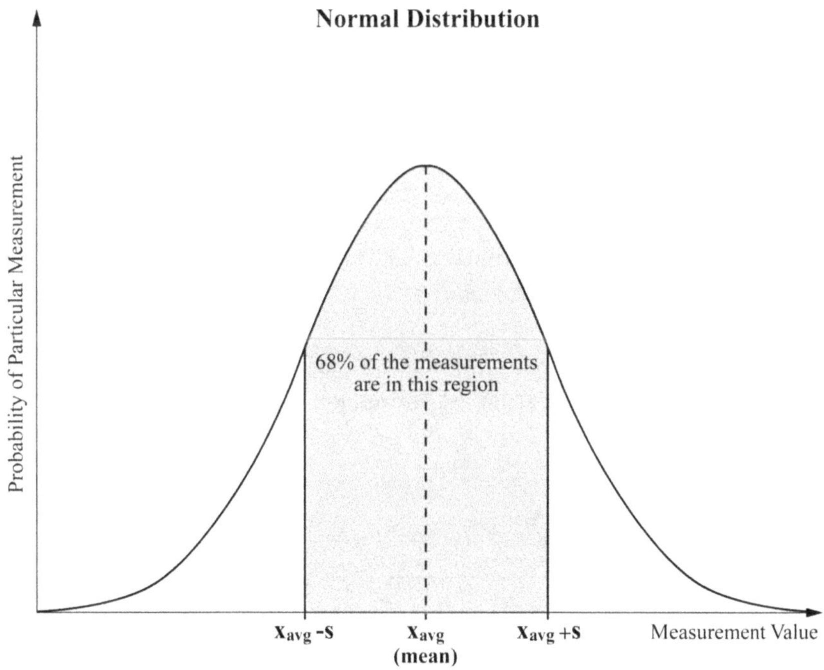

Figure 1. In a normal distribution, half the data are greater than the mean, x_{avg}, and half are less than the mean. The mean, the median, and the mode are the same number. A random measurement has a 68% probability of falling within one standard deviation of the mean.

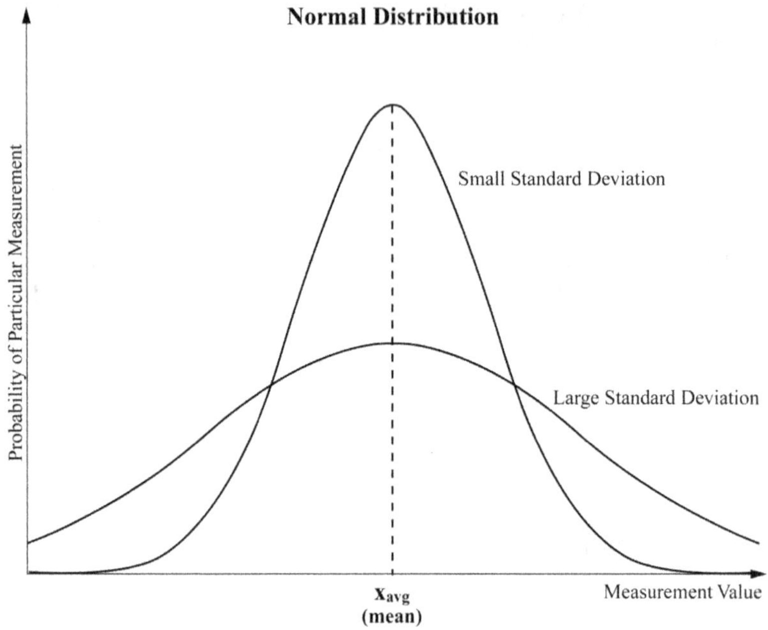

Figure 2. Data with a small standard deviation are grouped closely together. Data with a large standard deviation are more spread out. Both sets of data shown have the same mean, x_{avg}.

We now show how to find the standard deviation of a set of data. We can always compute it from the formula:

$$s = \sqrt{\frac{\sum_{i=1}^{n} (x_i - x_{avg})^2}{n-1}}$$

where

s = standard deviation
x_i = the data points
x_{avg} = the mean
n = number of data points

A more practical method to compute 's' is to use a calculator or a computer. There are too many slightly different calculator methods to list them all in detail here, so we must refer to the instruction manual that comes with the calculator. We give only a general procedure here.

A scientific calculator will usually have the statistics programs. The standard deviation may have the notation 's' or σ_{n-1}. As we enter the data into our calculator, we usually push the button labeled "data" or $\sum$ or $\sum$+. Once the data are entered, we push the 's' button to display the standard deviation. The calculator will also give us the mean (average) if we push the button labeled x´ , x_{ave}, or $\bar{x}$.

To use Microsoft Excel spreadsheet software to compute statistics, follow these steps:

- First enter the data into a column on a spreadsheet.
- To get the mean (average): in an empty cell type =AVERAGE(*select the data cells*) and hit enter.
- To get the sum: in an empty cell type =SUM(*select the data cells*) and hit enter.
- To get the median: in an empty cell type =MEDIAN(*select the data cells*) and hit enter
- To get the standard deviation: in an empty cell type =STDEV(*select the data cells*) and hit enter.

To let the computer know we are about to enter a formula, we must first enter an '=' sign. There are no spaces between the '=' sign and the function name or between the function name and the first parenthesis '('. To select the data cells, we can either click on the first cell and drag down to the last cell, or we can click on the first cell and hold the shift key as we click on the last cell. Cells can also be selected by typing in the range, first cell + colon + last cell, e.g., D3:D23. After the data cells are selected we must close the parentheses by typing ')'. The DFT data can serve as an example.

The following set of data has fewer points, so we can more easily check that we are performing the operations correctly.

Example: Find the average and the standard deviation of the following five DFT spot measurements taken over a 100 ft^2 area: 2.4, 2.7, 1.9, 2.2, and 2.5.

Solution: Use a calculator or a computer to get:

$x_{avg} = 2.34$ mean or average

$s = 0.272763$ standard deviation (can be rounded to 0.27)

In the example above, the average of the five spot measurements is the DFT according to SSPC PA 2. How confident are we that this average is close to the "true" DFT? How close are we to the "true" DFT? The "true" DFT can be considered the average of all the spot measurements taken on every conceivable spot on the surface.

Since we cannot measure every spot, we sample five spots and compute the average. If we sampled ten spots we would expect our average to be closer to the "true" value. The average of 100 spots would be even closer. The average of our five spots represents a sample mean.

We want to know how close this sample average is to the true DFT. The best we can do is establish a confidence interval. The 95% confidence interval is most commonly used. In other words, how big of an interval should we choose so that we are 95% confident that the true value lies in this interval. The formula, which can be found in statistics books, is that the true value can be found in the interval:

$$x_{avg} \pm t_{\alpha/2} \cdot \frac{s}{\sqrt{n}}$$

where

x_{avg} = sample mean

$t_{\alpha/2}$ = student t distribution coefficient (depends on confidence and n)

n = number of measurements in the sample

This formula assumes a normal distribution of data. The coefficient, $t_{\alpha/2}$, from the Student 't' distribution, can be found in a table and depends on the confidence desired (99%, 95%, 50%, etc.) and the number of data points in the sample. For five data points and 95% confidence, $t_{\alpha/2}$= 2.776; for 20 data points and 95% confidence, $t_{\alpha/2}$ = 2.093.

Example: What is the 95% confidence interval for the DFT data given in the previous example?

Solution: In the sample mean, x_{avg} = 2.34, there were five data points (n = 5), the standard deviation, s = 0.272763, and $t_{\alpha/2}$ = 2.776. (We use more significant figures for s in the intermediate computations.) Substituting into the formula we get:

$$x_{avg} \pm t_{\alpha/2} \cdot \frac{s}{\sqrt{n}} = (2.34) \pm (2.776)\frac{(0.272763)}{\sqrt{5}} = 2.34 \pm 0.3386$$

This can be rounded off to 2.34 ± 0.34. Thus, we are 95% confident that the true DFT lies somewhere between 2.34 – 0.34 and 2.34 + 0.34, or between 2.00 and 2.68 mils.

PART II – SPECIFIC COATINGS INDUSTRY APPLICATIONS

Part II of this book is a compilation of problems typically encountered in the coatings industry. Solving these problems assumes an understanding of the mathematical concepts covered in Part I.

Chapter 7 – Thinning, Spreading Rate, and DFT

Objective: To learn how to compute DFT, WFT, or percent volume solids if two of these three quantities are known; to compute theoretical coverage, with or without thinning; to compute transfer efficiency; and to estimate the amount of material needed for a job.

7.1 Volume Solids – Relating Thinning, Spreading Rate, and DFT

Paint is composed of pigment and vehicle. The pigment is a finely ground solid such as zinc dust, titanium dioxide, iron oxide, micaceous iron oxide, talc, etc. Pigment can be further subdivided into rust inhibitors, extenders, color enhancers, clays, and other additives. The vehicle is composed of resin (or binder) and solvent. The resin can be considered the glue that holds the pigment together after the solvent evaporates. Epoxy, polyurethane, and alkyd are examples of commonly used resins. After the solvent evaporates, the non-volatile vehicle (resin) remains. This resin, along with the pigment, comprise the paint solids. Even though the resin may be liquid in the can, it is considered part of the paint solids because it cures to a solid state.

The technical data sheet supplied by the coating manufacturer will state the percent solids of the coating. The solids may comprise a certain percentage of the weight or a percentage of the volume. These two numbers are not the same because the density of the solid portion is usually greater than the density of the solvent, which evaporates. The percent solids by volume is the quantity usually stated on the tech data sheet. We would have to know the density and the weight of each component to convert between percent by weight and percent by volume.

Often the manufacturer will provide both the percent of solids and the percent of pigment by weight. Percent of solids is the ratio of the weight of solids to the total weight of the paint, with this ratio expressed as a percent. Similarly, the weight percent of pigment is the ratio of the weight of pigment to the weight of paint. Both percent solids and percent pigment can be indicative of the quality of the paint. Excess solvent, over that needed for application, dilutes the paint and lowers the quality.

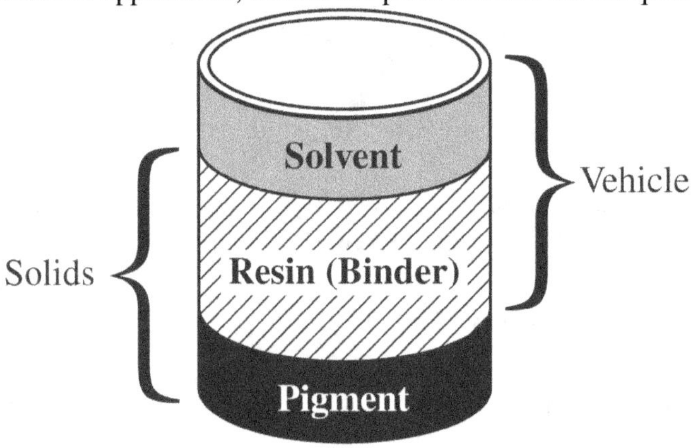

Figure 1. Paint is composed of pigment and vehicle. The solvent and resin (binder) are the components of the vehicle.

Example: Laboratory analysis determined that FlowRite coating had 78% total solids by weight and 36% pigment by weight. From these data, calculate the percent of binder in the vehicle portion of the coating.
Solution: Since the data are given as percentages, if we assume we have 100 lb (or 100 kg) of paint, the 78% total solids can be interpreted as 78 lb (78% of 100 lb = 78 lb). Let us assign letters to all quantities, knowns as well as unknowns, and then write the relationships as equations.

weight of total solids	= S	= 78 lb
weight of pigment	= P	= 36 lb
weight of binder (or resin)	= B	
weight of vehicle	= V	
weight of thinner (more precisely, volatile vehicle)	= T	

The question can be paraphrased into an "is" statement:

Binder is what percent of vehicle?

$$B = x \cdot V$$

where x is the percent. In order to find x, we must find B and V.

We first find B. Since total solids is comprised of pigment and binder, we have:

$$P + B = S$$

or

$$B = S - P$$

$$B = 78 \text{ lb} - 36 \text{ lb} = 42 \text{ lb}$$

Thus, there is 42 lb of binder in 100 lb of paint, or 42% of the paint is binder.

We next find V. Since 100% of the paint (or 100 lb) is comprised of pigment and vehicle, we have:

$$P + V = 100 \text{ lb}$$

$$V = 100 \text{ lb} - P$$

$$V = 100 \text{ lb} - 36 \text{ lb}$$

$$V = 64 \text{ lb}$$

Thus, there is 64 lb of vehicle in 100 lb of paint, or 64% of the paint is vehicle (binder and solvent).

Returning to our earlier relationship between binder and vehicle, and substituting, we get:

$$B = x \cdot V$$

$$42 \text{ lb} = x \cdot 64 \text{ lb}$$

Dividing both sides of the equation by 64 lb and reversing the equation we get:

$$x = \frac{42 \cancel{\text{lb}}}{64 \cancel{\text{lb}}} = 0.65625 = 66\%$$

Thus, binder is 66% of the vehicle. Since the vehicle is composed of binder and solvent, the other 34% of the vehicle is solvent (100% – 66% = 34%).

When we say a coating has 65% solids, we usually mean 65% by volume. In other words, if 65% of each gallon is solids, the other 35% must be solvent that evaporates after application (65% + 35% = 100%). If we left the lid off the can and let the paint harden, the dried paint would only fill 65% of the can. If, instead of leaving it in the can to dry, we spread the paint over the wall, the volume of dried paint on the wall is 65% of one gallon. Since volume is area times thickness, and since the area covered by the wet paint is the same as the area covered by the dry paint, the thickness of the dry paint must be 65% of the thickness of the wet paint. Figure 2 illustrates why this is so.

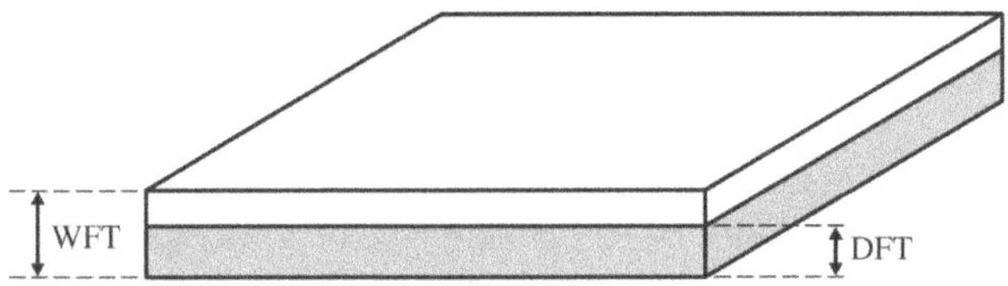

Figure 2. The relationship of DFT and WFT.

For this particular coating, we can state that the dry film thickness (DFT) is 65% of the wet film thickness (WFT). This statement can be translated into a mathematical equation where we let the single letter D = DFT and W = WFT. The percent is expressed as a decimal.

DFT	is	65%	of	WFT
D	=	0.65	·	W

This equation has two letters, so if we know one of the letters, we can solve for the other one.

Example: A paint with 65% volume solids is applied at 5.0 mils wet. What will be the DFT?
Solution: We first find a relationship (equation) that relates what we know and what we do not know. When we substitute for the known quantities, we substitute the units (dimensions) as well as the numerical value for the given letter in the equation.

$$D = 0.65 \cdot W$$
$$D = 0.65 \cdot (5.0 \text{ mils})$$
$$D = 3.25 \text{ mils}$$

We round the answer to two significant figures and state that the DFT will be 3.2 mils.

Example: What WFT is needed to get 4.0 mils DFT if the paint is 65% solids by volume?
Solution: We know DFT and percent solids, and we wish to find WFT. We state the relationship in a sentence, translate the sentence into an equation, and then substitute into the equation, which relates the known quantities and the unknown quantity.

The DFT is 65% of the WFT.

$$D = 0.65 \cdot W$$

$$4.0 \text{ mils} = 0.65 \cdot W$$

We reverse the equation and divide both sides by 0.65.

$$0.65 \cdot W = 4.0 \text{ mils}$$

$$\frac{0.65 \cdot W}{0.65} = \frac{4.0 \text{ mils}}{0.65}$$

$$W = 6.1538 \text{ mils}$$

Rounding to two significant figures we determine that a WFT of 6.2 mils is required to get a DFT of 4.0 mils.

When solving computation problems it is good practice to always ask ourselves if the answer is reasonable. In these two examples we would ask, “Is the WFT more than the DFT?” This would not be the case if we mixed up our multiplication and division.

Another question we would ask ourselves is, “Are these ballpark values?” If we get 62 mils for a WFT, there is obviously something wrong since this would clearly be too thick; perhaps the decimal point did not record when we entered the number on our calculator. If we get 0.062 mils as our answer, we probably forgot to convert the percent to a decimal.

Generalizing the previous examples, we derive the basic relationship among DFT, WFT, and percent volume solids. For a coating that is 100% solids, WFT = DFT, since 100% = 1.

Formula Relating DFT, WFT, and Percent Volume Solids

$$\text{DFT} = \left(\begin{array}{c}\text{percent volume solids}\\ \text{expressed as a decimal or a fraction}\end{array}\right) \cdot \text{WFT}$$

or using single letters

☞ $$D = P_S \cdot W$$

The equation can be rearranged to solve for WFT

☞ $$W = \frac{D}{P_S}$$

The equation to solve for percent volume solids is

☞ $$P_S = \frac{D}{W}$$

where

D = dry film thickness
W = wet film thickness
P_S = percent volume solids (expressed as a decimal or a fraction)

7.2 Computing Theoretical Coverage

If we know the area to be covered, the target DFT, and the volume solids of the paint, how many gallons do we need? This is a common estimating problem. The "theoretical" coverage will give the volume of paint needed if there are no losses due to overspray, spillage, paint flushed from the lines, etc. In practice, we will always need more paint than the theoretical computation predicts, but the theoretical computation gives us a good starting point.

There are several approaches to these problems, one of which is to remember a formula and plug in the values. If we do not have our formula handy, we can always approach the problem as a conversion of units problem. We will use this approach to develop a formula. First, we will do a few numerical examples.

Example: How much area can be covered by one gallon of paint if the wet film thickness is 5.0 mils?
Solution: We know (from Section 3.3) that one gallon will cover 1604 ft^2 at a thickness of one mil, or that 1 gal = 1604 mil·ft^2. From geometry we

know that volume equals area times thickness, or:

$$V = A \cdot T$$

where V = volume, A = area, and T = thickness. Rearranging the equation (divide both sides by T) to solve for area we get:

$$A = \frac{V}{T}$$

Substituting for V = 1 gal = 1604 mil·ft^2, and T = 5.0 mils we get:

$$A = \frac{1604\ \cancel{mil} \cdot ft^2}{5.0\ \cancel{mil}}$$
$$A = 320.8\ ft^2$$

Rounding to two significant figures, we state that one gallon will cover 320 ft^2 at a wet film thickness of 5.0 mils.

Example: How many gallons (theoretically) are needed to apply a 4.0 mil wet coat to 6500 ft^2?
Solution: We know area and thickness and we wish to find volume. Substituting into the formula that relates the knowns to the unknowns we get:

$$V = A \cdot T$$
$$V = (6500\ ft^2) \cdot (4.0\ mils)$$
$$V = 26{,}000\ mil \cdot ft^2$$

The next step is to convert mil · ft^2 into gallons using the conversion factor 1 gal = 1604 mil · ft^2

$$V = \frac{26{,}000\ \cancel{mil \cdot ft^2}}{1} \cdot \frac{1\ gal}{1604\ \cancel{mil \cdot ft^2}} = 16.2094\ gal$$

So it would take 16.2 gal to cover the 6500 ft^2 at a WFT of 4.0 mils.

Example: How much paint (theoretically) is needed to cover 5000 ft^2 at a dry film thickness of 2.0 mils if the volume solids of the paint is 72%?
Solution: In this problem we must allow for the solvent that evaporates. There are several ways to approach this. One method is to determine the wet film thickness required to achieve 2.0 mils dry, and then compute the volume needed as was done in the previous example. We first find the

WFT required to give 2.0 mils dry.

$$WFT = \frac{DFT}{\%\ solids}$$

OR

$$W = \frac{D}{P_S}$$

$$W = \frac{2.0\ mils}{0.72}$$

$$W = 2.77778$$

We will carry an extra significant figure through the intermediate calculations and let WFT = 2.78 mils. If we do this problem on a calculator, we do not have to round and can carry all the figures. Now that we know the WFT, we compute the volume as done above and convert to gallons. Using volume (V) equals area (A) times thickness (T), where T is the WFT:

$$V = A \cdot T$$

$$V = (5000\ \cancel{ft^2}) \cdot (2.78\ \cancel{mils}) \cdot \frac{1\ gal}{1604\ \cancel{mil} \cdot \cancel{ft^2}}$$

$$V = 8.6658\ gal$$

Rounding the answer we state that the job will theoretically require 8.7 gal.

Using these few examples as models, we can develop a general formula.

$$\text{Volume of paint needed} = (\text{area in ft}^2) \cdot (\text{WFT in mils}) \cdot \frac{1\ gal}{1604\ mil \cdot ft^2}$$

Replacing WFT with $\left[\frac{DFT}{\%\ volume\ solids}\right]$ we get

$$\text{Volume} = (\text{area in ft}^2) \cdot \frac{\text{DFT in mils}}{\%\ \text{volume solids}} \cdot \frac{1\ gal}{1604\ mil \cdot ft^2}$$

The last fraction is equal to 1, and converts from mil $\cdot$ ft^2 to gallons.

We can write the formula using letters as $V = A \cdot \frac{D}{P_S} \cdot \frac{1}{1604}$ or $V = \frac{A \cdot D}{P_S \cdot 1604}$

Once a formula has been developed, it is not necessary to start from first principles for each problem. We will use this formula in the next example.

Formula for Volume of Paint Needed

$$V = \frac{A \cdot D}{1604 \cdot P_S}$$

where

V = volume in gallons
A = area in ft^2
D = dry film thickness in mils
P_S = percent volume solids (expressed as a decimal or a fraction)
1604 = the conversion from mil · ft^2 to gal

Example: How many gallons are needed (theoretically) to cover 15,000 ft^2 at a dry film thickness of 3.5 mils if the coating is 82% solids by volume?

Solution: We first identify known quantities and assign letters to them: Area A = 15,000 ft^2; dry film thickness D = 3.5 mils; percent volume solids P_s = 82% = 0.82. The unknown quantity is volume V. Substituting into the equation above that relates the knowns to the unknowns we get:

$$V = \frac{A \cdot D}{1604 \cdot P_S} \cdot \frac{(15{,}000) \cdot (3.5)}{1604 \cdot (0.82)} = 39.915$$

Rounding the answer we state that it will theoretically take 40 gal to do the job.

Alternative solution: Another approach to this problem is to first determine how much area one gallon will cover (find ft^2/gal), and then divide this number into the total area to get the number of gallons needed.

We first determine the coverage per gallon. Since the paint is only 82% solids it will only cover 82% of 1604 ft^2 at one mil dry.

Now 82% of 1604 is (0.82) · (1604) = 1315. So one gallon will cover 1315 ft² at 1 mil DFT. Since volume equals area times thickness, V = A · T, we can solve for A to get A = V/T. Substituting into the formula:

$$A = \frac{V}{T}$$

$$A = \frac{1315\,\cancel{\text{mil}} \cdot \text{ft}^2}{3.5\,\cancel{\text{mil}}}$$

$$A = 376\ \text{ft}^2$$

So 1 gal will cover 376 ft² at 3.5 mils DFT.

Next we have to find the number of gallons to cover 15,000 ft². We get this number by dividing 15,000 by 376:

$$\frac{15{,}00\,\cancel{\text{ft}^2}}{376\frac{\cancel{\text{ft}^2}}{\text{gal}}} = 39.89\ \text{gal}$$

Notice that the units work out properly. If the units do not work out, we definitely have an incorrect formula or we have substituted incorrectly. Looking only at the units:

$$\frac{\text{ft}^2}{\frac{\text{ft}^2}{\text{gal}}} = \frac{\cancel{\text{ft}^2}}{1} \cdot \frac{\text{gal}}{\cancel{\text{ft}^2}} = \text{gal}$$

So it will take about 40 gal, the same as we determined by the first method.

7.3 Effect of Thinning on Spreading Rate (Coverage)

If we add thinner to the paint, we will change the percent volume solids. This in turn will change the relationship between DFT and WFT as well as affect the coverage area. If we can compute the percent solids after thinning, we can then use this new percentage in the formulas given in the previous section. The percent solids, P_s, is defined as the volume of solids divided by the total volume of the paint. This gives a fraction that can also be written as a decimal or a percent. If we add thinner, we add nothing to the solids but we increase the volume of the paint. Thinking of percent solids as a fraction, adding thinner increases the denominator but does nothing to the numerator, which then decreases the value of P_s.

Example: Covermore paint is 78% solids by volume. If we thin it by adding one pint per gallon, what is the percent volume solids of the thinned paint?
Solution: We must express the volume quantities in the same units. One pint is 1/8 gal or 0.125 gal. In one gallon of unthinned paint there will be 0.78 gal of solids. There will also be 0.78 gal of solids in the thinned paint, which will now have a volume of one gallon plus one pint or 1.125 gal. The percent solids of the thinned paint is then:

$$P_S = \frac{0.78 \text{ gal}}{1 \text{ gal} + 0.125 \text{ gal}}$$

$$P_S = \frac{0.78 \not{\text{gal}}}{1.125 \not{\text{gal}}}$$

$$P_S = 0.69333$$

Rounding to two significant figures, $P_S = 0.69$ or 69%. So thinning one pint per gallon reduced the volume solids from 78% to 69%.

Some tech data sheets state that thinning is allowable up to so many percent, such as 5%. This means that the maximum amount of thinner that can be added to a gallon of paint is 5% of one gallon or 0.05 gal.

Example: Quickcoat is 86% solids and allows thinning up to 5% by volume. What is the percent volume solids at the maximum allowable thinning?
Solution: Start with one gallon. Both before and after thinning the volume occupied by the solids is 86% of one gallon or 0.86 gal. The volume of the thinned paint has increased by 5% of one gallon or 0.05 gal. We find the percent solids by dividing the volume of solids by the total volume.

$$P_S = \frac{0.86 \not{\text{gal}}}{1.05 \not{\text{gal}}}$$

$$P_S = 0.81904$$

Round to get $P_S = 0.82$, so the volume solids of the thinned paint is 82%.

Example: How much thinner should we add to a 5-gal bucket of paint to thin it by 12%?
Solution: Paraphrasing the problem into an "is" statement we get, "The amount of thinner is 12% of 5 gal." Let V_T be the volume of thinner.

Translate the statement into an equation.

$$\text{Volume of thinner is } 12\% \text{ of } 5 \text{ gal}$$
$$V_T = 0.12 \cdot 5 \text{ gal}$$

Doing the multiplication we get $V_T = 0.6$ gal. Since there are 8 pints in one gallon, we can convert 0.6 gal to pints:

$$\frac{0.6 \cancel{\text{gal}}}{1} \cdot \frac{8 \text{ pints}}{1 \cancel{\text{gal}}} = 4.8 \text{ pints}$$

So we could add 5 pints of thinner to a 5-gal bucket of paint.

If we want to estimate how many gallons are needed for a job, we compute the theoretical coverage based on the unthinned paint. **The number of gallons of paint we need to purchase is the same whether or not we thin the paint.** However, once we thin the paint, the WFT must increase to compensate for the additional thinner.

Example: Lusterlux is 58% solids by volume. If it is thinned by 10%, what WFT is needed to give a DFT of 1.5 mils?
Solution: We first compute the volume solids of the thinned paint from the formula:

$$\begin{array}{c}\text{percent volume solids}\\ \text{(expressed as a decimal)}\end{array} = \frac{\text{volume of solids}}{\text{original volume + volume of thinner}}$$

Substituting into the formula:

$$P_S = \frac{0.58 \text{ gal}}{1 \text{ gal} + 0.10 \text{ gal}}$$
$$P_S = \frac{0.58 \cancel{\text{gal}}}{1.10 \cancel{\text{gal}}}$$
$$P_S = 0.5272$$

The volume solids of the thinned paint is 52.7%. (We carry an extra significant figure in the intermediate computations.)

We use this to compute the WFT by substituting into the formula:

$$\text{WFT} = \frac{\text{DFT}}{\text{percent volume solids}}$$

$$W = \frac{D}{P_S}$$

$$W = \frac{1.5 \text{ mils}}{0.527}$$

$$W = 2.846 \text{ mils}$$

So we need to apply the thinned paint at a wet thickness of 2.8 mils to get 1.5 mils dry.

These last two formulas can be combined. First we assign letters to the quantities to aid in the algebraic manipulation and make the formulas look simpler. Let:

V_o = original volume of paint (usually 1 gal)
V_T = volume of thinner added
V_S = volume of solids of unthinned paint from tech data sheet (can be measured in gal)
D = dry film thickness
W = wet film thickness
P_S = percent volume solids of the thinned paint (expressed as a decimal or a fraction)
P_o = original percent volume solids (unthinned) (expressed as a decimal or a fraction)

Then:

$$V_S = P_o \cdot V_o$$

If $V_o = 1$, then V_S is numerically equal to P_o

Also:

$$P_S = \frac{V_S}{V_o + V_T}$$

Since

$$D = P_S \cdot W$$

We can substitute for P_s to get the dry film thickness:

$$D = \left[\frac{V_S}{V_o + V_T}\right] \cdot W$$

If we know the dry film thickness and wish to find the wet film thickness, the formula can be rearranged to give:

$$W = \frac{D(V_o + V_T)}{V_S}$$

Remember that if $V_o = 1$, V_s is numerically equal to the percent volume solids of the unthinned paint, P_o.

Let us repeat the previous example using the formula.

Example: Lusterlux is 58% solids by volume. If it is thinned by 10%, what WFT is needed to give a DFT of 1.5 mils?
Solution: We let $V_0 = 1$ gal and assign letters to the known quantities:

$P_o = 58\% = 0.58$
$V_T = 10\%$ of 1 gal $= 0.10$ gal
$D = 1.5$ mils

The unknown quantity is the WFT, which we call W. We now substitute into the formula:

$$W = \frac{D \cdot (V + V_T)}{V_S}$$

$$W = \frac{1.5 \text{ mils} \cdot (1 \text{ gal} + 0.10 \text{ gal})}{0.58 \text{ gal}}$$

$$W = \frac{1.5 \text{ mils} \cdot (1.10 \cancel{\text{gal}})}{0.58 \cancel{\text{gal}}}$$

$$W = 2.8448 \text{ mils}$$

So after thinning , we need to apply a wet film of 2.8 mils.

Relationship Among DFT, WFT, Percent Volume Solids, and Volume of Thinner

☞ To find DFT use

$$D = \left[\frac{V_S}{V_o + V_T}\right] \cdot W$$

☞ To find WFT use

$$W = \frac{D(V_o + V_T)}{V_S}$$

☞ To find V_S use

$$V_S = \frac{D(V_o + V_T)}{W}$$

where

V_O = original volume of paint (usually 1 gal)
V_T = volume of thinner added
V_S = volume of solids of unthinned paint from tech data sheet (can be measured in gal)
D = dry film thickness
W = wet film thickness

☞ If $V_O = 1$, V_S is numerically equal to the percent volume solids (in decimal form) of the unthinned paint.

☞ Corresponding formulas using **percent** thinner are given later in this section.

Example: The tech data sheet for SpreadRite states that after thinning by 10%, a wet film thickness of 5.0 mils will result in a dry film thickness of 3.5 mils. What is the percent volume solids of SpreadRite?
Solution: We let $V_o = 1$ gallon and assign letters to the known quantities:

W = 5.0 mils
D = 3.5 mils

The volume of thinner, V_T, is 10% of 1 gal or $V_T = 0.10$ gallon.

Substituting into the formula above we get:

$$V_S = \frac{D(V_o + V_T)}{W}$$

$$V_S = \frac{3.5\,\cancel{\text{mils}} \cdot (1\text{ gal} + 0.1\text{ gal})}{5.0\,\cancel{\text{mils}}}$$

$$V_S = 0.77\text{ gal}$$

Since there is 0.77 gal of solids in 1 gal (V_o = 1 gal), the percent volume solids, P_s, is 0.77 or 77%.

We can combine the formulas relating area, volume, DFT, and WFT with the formulas for thinning to get a formula for spreading rate of thinned paint. The formula

$$V = \frac{A \cdot D}{1604 \cdot P_S}$$

can be rearranged to give spreading rate (area per unit volume).

$$\frac{A}{V} = \frac{1604 \cdot P_S}{D}$$

where

V = volume in gal
A = area in ft^2
D = dry film thickness in mils
P_s = percent volume solids after thinning (expressed as a decimal or a fraction)
1604 = the conversion from mil·ft^2 to gal

We also have:

$$P_S = \frac{V_S}{V_o + V_T}$$

where

V_o = original volume of paint (usually 1 gal)

V_T = volume of thinner added
V_S = volume of solids of unthinned paint from tech data sheet (can be measured in gal)
P_S = percent volume solids of the thinned paint (expressed as a decimal or a fraction)

If we divide the numerator and the denominator of the fraction on the right side of the equation by V_o, we get:

$$P_S = \frac{V_S}{V_o + V_T} = \frac{\frac{V_S}{V_o}}{\frac{V_o}{V_o} + \frac{V_T}{V_o}}$$

$$P_S = \frac{P_o}{1 + P_T}$$

since

$$\frac{V_S}{V_o} = P_o \quad \text{and} \quad \frac{V_T}{V_o} = P_T$$

Substituting this expression for P_S into the formula for spreading rate we get:

$$\frac{A}{V} = \frac{1604 \cdot P_S}{D} = \frac{1604 \cdot \left[\frac{P_o}{1 + P_T}\right]}{D} = \frac{1604 \cdot P_o}{D \cdot (1 + P_T)}$$

Example: Laboratory analysis determined that WonderCoat contained 27% solvent by volume. If it is thinned by 10%, what is the theoretical spreading rate at 3.5 mils DFT?

Solution: The spreading rate formulas are concerned with volume solids, but we are given that 27% of the volume is solvent. The volume solids is then 100% – 27%, or 73%. We assign letters to the known quantities:

$P_o = 73\% = 0.73$	percent volume solids of unthinned paint
$P_T = 10\% = 0.10$	percent thinner added
$D = 3.5$ mils	DFT

Formula for Computing Percent Volume Solids After Thinning

$$P_S = \frac{P_o}{1 + P_T}$$

where

P_S = % volume solids after thinning (expressed as a decimal or a fraction)

P_O = original percent volume solids (unthinned) (expressed as a decimal or a fraction)

P_T = percent of thinner by volume added

Formula for Spreading Rate (A/V) of Thinned Paint

$$\frac{A}{V} = \frac{1604 \cdot P_o}{D \cdot (1 + P_T)}$$

where

A = area in ft^2

V = volume in gal

D = dry film thickness in mils

P_o = percent volume solids of unthinned paint (expressed as a decimal or a fraction)

P_T = percent of thinner by volume added (expressed as a decimal or a fraction)

1604 = the conversion from $mil \cdot ft^2$ to gal

We find the theoretical spreading rate by substituting these values into the formula:

$$\frac{A}{V} = \frac{1604 \cdot P_o}{D \cdot (1 + P_T)} = \frac{1604 \cdot (0.73)}{(3.5) \cdot (1 + 0.10)} = 304.135$$

So after thinning by 10%, the theoretical spreading rate of WonderCoat is about 300 ft^2/gal.

One of the more frequent decisions we must make is to determine the wet film thickness needed to obtain the specified dry film thickness if the paint is thinned by so many percent. In an earlier example, we first computed the volume of thinner needed and then computed the required WFT. This extra step is not necessary if we can use a formula containing percents.

Recall that

$$\text{wet film thickness} = \frac{\text{dry film thickness}}{\text{percent volume solids}} \quad \text{OR} \quad W = \frac{D}{P_S}$$

AND

$$P_S = \frac{P_o}{1 + P_T}$$

If we substitute the second formula into the first one we get:

$$W = \frac{D}{P_S} = \frac{D}{\frac{P_o}{1 + P_T}} = D \cdot \frac{1 + P_T}{P_o}$$

OR

$$W = \frac{D \cdot (1 + P_T)}{P_o}$$

where

W = wet film thickness
D = dry film thickness
P_T = percent thinner added
P_o = percent volume solids of unthinned paint (expressed as a decimal or a fraction)

We can use this formula to solve an earlier example.

Example: Lusterlux is 58% solids by volume. If it is thinned by 10%, what WFT is needed to give a DFT of 1.5 mils?
Solution: We identify our known quantities and assign letters to them. Percent solids P_S = 58% = 0.58, percent of thinner added is P_T = 10% = 0.10, and DFT is D = 1.5 mils. Our unknown is wet film thickness, W. Substituting these numbers into the formula for wet film thickness we get:

$$W = \frac{D \cdot (1 + P_T)}{P_o}$$

$$W = \frac{(1.5 \text{ mils}) \cdot (1 + 0.10)}{0.58}$$

$$W = 2.8448 \text{ mils}$$

Rounding to two significant figures we can state that we need 2.8 mils wet to achieve 1.5 mils dry.

This formula was easier to use than the previous one because the data was given as percents. The previous formula would be easier if the data had been given as a particular *volume* of thinner allowed. The formula for wet film thickness in terms of percent of thinner can be rearranged to find the other quantities. See summary of formulas on page 117.

7.4 Effect of Thinning on VOC Content

When thinner (other than water) is added to paint, the concentration of volatile organic compounds (VOC) will increase. In so doing, we must be sure we do not exceed the legal limit for VOC content. The following examples (metric and US customary) will illustrate methods for checking VOC content after thinning. The VOC content of the paint as received from the manufacturer should be given on the tech data sheet. If the quoted VOC content exceeds the local limit, we must find another product. VOC content is usually expressed as so many pounds per gallon or grams per liter.

Example: The tech data sheet gives the VOC concentration of our coating as 2.8 lb/gal. The local regulation lists the upper limit as 360 g/L. Are we in compliance?
Solution: In order to compare two quantities, they must both be expressed in the same units. Hence we must convert lb/gal into g/L or vice versa. To

Relationship Among DFT, WFT, Percent Volume Solids, and Percent of Thinner

☞ To find DFT use

$$D = \left[\frac{P_S}{1+P_T}\right] \cdot W$$

☞ To find WFT use

$$W = \frac{D(1+P_T)}{P_S}$$

☞ To find P_S use

$$P_S = \frac{D(1+P_T)}{W}$$

where

P_T = percent of thinner by volume added (expressed as a decimal or a fraction)

P_S = percent volume of solids of unthinned paint (expressed as a decimal or a fraction)

D = dry film thickness

W = wet film thickness

☞ Corresponding formulas using **volume** of thinner were given previously in this section.

do this conversion, we must look up the equivalent values (see Appendix D), 1 lb/gal = 120 g/L. (A more exact equivalence is 1 lb/gal = 119.83 g/L.) We will convert 2.8 lb/gal into g/L.

$$\frac{2.8\frac{\cancel{lb}}{\cancel{gal}}}{1} \cdot \frac{120\frac{g}{L}}{1\frac{\cancel{lb}}{\cancel{gal}}} = 336\frac{g}{L}$$

Since 336 is less than 360, the coating meets the VOC regulation.

The next question is, "How much thinner can we add to a gallon of paint and still conform to the regulation?" This is not as straight forward to answer, because adding thinner increases the weight of VOCs and at the same time increases the volume. The VOC limit is expressed as weight per unit volume. To make an exact computation we must know the density (or specific gravity) of the thinner. Like most chemists, we will use the metric system. We must look in a handbook or product data sheet to find that the thinner in this example has a density of 0.81 g/mL or 810 g/L. We can also assume that this thinner is 100% VOCs. Let the letter, w, represent the weight (in grams) of additional thinner that we add. Density is defined as weight divided by volume, or D = w/V. Rearranging the formula we get w = V · D. The weight, w, of added thinner is then w = $V_T \cdot 810$ where w is in grams and the volume of thinner, V_T, is in liters.

We start with 1 L of paint. After adding thinner, the volume is (1 + V_T) in liters. As computed above, one liter of paint contains 336 grams of VOCs. After thinning, the paint will contain (336 + w) grams of VOCs. We form a proportion by setting the weight to volume rate after thinning to the weight to volume rate allowed.

$$\frac{336 + w}{1 + V_T} = \frac{360}{1}$$

To solve an equation, there can only be one variable. We can replace w with (810 · V_T) to get:

$$\frac{336 + 810 \cdot V_T}{1 + V_T} = \frac{360}{1}$$

The equation now has only the letter, V_T. If we cross multiply and simplify the equation we get:

$$(336 + 810 \cdot V_T) \cdot 1 = 360 \cdot (1 + V_T)$$
$$336 + 810V_T = 360 + 360V_T$$
$$810V_T - 360V_T = 360 - 336$$
$$450V_T = 24$$
$$V_T = \frac{24}{450}$$
$$V_T = 0.053$$

So we can add 0.053 L or 53 mL of solvent to each liter of paint and still meet the VOC regulation.

How much can we add to each gallon? The ratio of volume of

additional solvent to original paint is:

$$\frac{0.053\ \text{L}}{1\ \text{L}} = 0.053$$

This ratio is the same whether we use gallons or liters. So, to one gallon of paint we can add 0.053 gallons of solvent. Using the fact that 128 oz = 1 gal, we can convert to ounces.

$$\frac{0.053\ \cancel{\text{gal}}}{1} \cdot \frac{128\ \text{oz}}{1\ \cancel{\text{gal}}} = 6.784\ \text{oz}$$

So we can add 6.8 oz of thinner to each gallon of paint and not violate the VOC regulation.

7.5 Computing DFT from Spreading Rate

In this section we are going to develop a method of computing the DFT instead of measuring it directly. Spreading rate is usually given as so many square feet per gallon of coating at a specified DFT. The same information could be conveyed by using the reciprocal of this number, but the units would be gallons per square foot and the value would be a decimal less than one. If we know the volume of paint used, the percent solids, and the area, we can compute the DFT. Earlier we derived the formula:

$$V = \frac{A \cdot D}{1604 \cdot P_S}$$

where

V = volume in gal
A = area in ft^2
D = dry film thickness in mils
P_S = percent volume solids
(expressed as a decimal or a fraction)
1604 = the conversion from $mil \cdot ft^2$ to gal

If we know V, A, and P_s, we can rearrange the equation to solve for D. We first multiply both sides of the equation by $1604 \cdot P_s$

$$V \cdot 1604 \cdot P_S = \frac{A \cdot D}{1604 \cdot P_S} \cdot 1604 \cdot P_S$$

After reordering the left side and canceling the $1604 \cdot P_s$ on the right side we get:

$$1604 \cdot V \cdot P_S = A \cdot D$$

Next divide both sides by A:

$$\frac{1604 \cdot V \cdot P_S}{A} = \frac{A \cdot D}{A}$$

$$\frac{1604 \cdot V \cdot P_S}{A} = D$$

Reverse the equation to get:

$$D = \frac{1604 \cdot V \cdot P_S}{A}$$

Example: A circular tank with diameter of 36 ft just had its floor coated with Toughbottom. The inspector forgot to measure the DFT of the floor before the tank was put back into service. Fortunately, the trash had not yet been hauled away and they counted seven empty 5-gal cans of Toughbottom. This is a flow coating so we can assume 100% of the coating is on the floor. Toughbottom is 92% solids by volume. Estimate the DFT.
Solution: We first identify the known quantities and assign letters to them.

tank diameter	d = 36 ft
volume of paint used	V = 35 gal (7 cans x 5 gal/can)
percent volume solids	P_s = 92% = 0.92

Next we assign letters to our unknown quantities:

D = DFT
A = area of tank bottom

We find a formula relating our knowns and our unknowns:

$$D = \frac{1604 \cdot V \cdot P_S}{A}$$

This formula has two unknown quantities (D and A), and we cannot solve an equation unless there is only one unknown letter. Fortunately, we can calculate the floor area using $A = \pi r^2$, where the radius, r, is one half the diameter, d, which we know to be 36 ft. Hence, r = 18 ft and area is:

$$A = \pi\,(18\ \text{ft})^2 = 1018\ \text{ft}^2$$

Next we substitute into the equation for D:

$$D = \frac{1604 \cdot (35) \cdot (0.92)}{(1018)}$$

$$D = 50.7$$

Thus, the calculated DFT of the floor coating is 51 mils.

7.6 Transfer Efficiency

The transfer efficiency is a measure of the percentage of coating transferred from the can to the surface being coated. For example, if we spray a 5 gal can of paint and only 4 gal is deposited on the surface being painted, the transfer efficiency, E_T, is the ratio 4 gal to 5 gal.

$$E_T = \frac{4\ \cancel{\text{gal}}}{5\ \cancel{\text{gal}}}$$

$$E_T = 0.8$$

Multiplying by 100% (which is equal to 1) we can express 0.8 as 80%. So the transfer efficiency is 80%. The other 20% is lost as overspray, left in the container, left in the hoses, etc.

Instead of using volume, transfer efficiency can be computed from the ratio of the mass or weight of coating on the surface to the total mass or weight sprayed.

Transfer efficiency can be defined by the formula:

$$\text{transfer efficiency} = \frac{\text{volume of coating applied to surface}}{\text{volume of coating used}}$$

OR

$$\text{transfer efficiency} = \frac{\text{mass (weight) of coating applied to surface}}{\text{mass (weight) of coating used}}$$

Example: 5 gal of Toughfloor was applied to a rectangular concrete pad 15 ft by 25 ft. The volume solids of the coating was 84%. The dry film thickness of the applied coating was measured to be 14 mils. What was the transfer efficiency?

Solution: We can get transfer efficiency if we can compare the volume of coating on the floor to the volume of coating used. Since the solvent has evaporated, only the solids are now on the floor.

Approach One – Compute the volume of solids on the floor and compare it to the volume solids in 5 gal. To get the volume of coating on the floor we use the general formula that volume equals area times thickness, and convert to gallons. The area of the rectangular floor is length times width or (25 ft) · (15 ft) = 375 ft^2.

$$\text{Volume} = \text{Area} \cdot \text{Thickness}$$

$$V = (375\ \text{ft}^2) \cdot (14\ \text{mils})$$

$$V = 5250\ \text{mil} \cdot \text{ft}^2$$

converting

$$V = 5250\ \cancel{\text{mil}} \cdot \cancel{\text{ft}^2} \cdot \frac{1\ \text{gal}}{1604\ \cancel{\text{mil}} \cdot \cancel{\text{ft}^2}}$$

$$V = 3.273\ \text{gal}$$

The volume of solid coating on the floor is 3.27 gal.

Now we must find the volume of solids in the 5 gal. We know that the volume of solids we started with is 84% of 5 gal, or (0.84) · (5 gal) = 4.2 gal.

Transfer efficiency is then the ratio:

$$E_T = \frac{3.27 \cancel{gal}}{4.2 \cancel{gal}}$$

$$E_T = 0.77857$$

Rounding and then converting to percent gives the transfer efficiency of this job as $E_T = 0.78$ or 78%.

Approach Two – We compute the WFT that would be required to achieve 14 mils dry. From this, compute the wet volume on the surface (in gallons), and finally compare to the 5 gal. Since WFT equals DFT divided by percent solids, we get:

$$W = \frac{14 \text{ mils}}{0.84} = 16.67 \text{ mils}$$

A WFT of 16.67 mils will give a DFT of 14 mils. The volume of paint 16.67 mils thick over an area of 375 ft² is:

$$V = A \cdot T$$

$$V = (375 \text{ ft}^2) \cdot (16.67 \text{ mils})$$

$$V = 6251 \text{ mil} \cdot \text{ft}^2$$

converting

$$V = 6251 \cancel{mil} \cdot \cancel{ft^2} \cdot \frac{1 \text{ gal}}{1604 \cancel{mil} \cdot \cancel{ft^2}}$$

$$V = 3.897 \text{ gal}$$

Thus, about 3.90 gal of liquid coating made it onto the floor. The transfer efficiency is then:

$$E_T = \frac{3.90 \cancel{gal}}{5.0 \cancel{gal}}$$

$$E_T = 0.78$$

Converting to percent, we again find that transfer efficiency is 78%.

Approach Three – We compute the expected DFT if all 5 gal made it to the floor and then compare to the actual DFT. Using the equation from the previous section to solve for D, we get:

$$D = \frac{1604 \cdot P_S \cdot V}{A}$$

$$D = \frac{1604(0.84)(5)}{375}$$

$$D = 17.96 \text{ mils}$$

If all 5 gal made it to the floor, the DFT would have been 17.96 mils. The transfer efficiency is then the ratio of 14 mils to 17.96 mils.

$$E_T = \frac{14 \cancel{\text{mils}}}{17.96 \cancel{\text{mils}}}$$

$$E_T = 0.779$$

Rounding and converting to a percent again gives a transfer efficiency of 78%.

7.7 Estimating Materials Needed

This section will tie together many of the concepts discussed so far. When estimating, old records and experience often help to fine tune estimates to make them more accurate.

Estimates on coating losses may range from 5% for a large flat surface to more than 40% for an open lattice. After theoretical quantities are calculated, other factors are taken into account.

The amount of paint used for a job is always greater than the theoretical amount calculated. Whether we brush, roll, or spray, there will be some paint left in the can, in the brush, roller, hoses, gun, etc., and some will fall on the ground as drips, splashes, and overspray.

In addition, the paint film is not perfectly uniform because of overlaps during application, which results in extra paint being used. One rule of thumb for calculating the additional amount of paint to order when painting a flat surface is given in Table 7.1.

Table 7.1

Amount of Paint to be Ordered Beyond the Theoretical Value When Painting a Flat Surface

APPLICATION METHOD	ADDITIONAL PAINT NEEDED
brush	5%
roll	10%
spray	20 – 30%

Example: We are preparing to order supplies for a job blasting and painting the exterior of a covered cylindrical fuel oil storage tank 50 ft in diameter and 28 ft high. The specification calls for a commercial blast (SSPC-SP 6) with a 2–3 mil anchor pattern. From the condition of the tank, we estimate that if we use #8 nozzles at 100 psi it will take 4.0 pounds of coal slag abrasive per square foot. The coatings are:

Primer	Inhibitive Epoxy; 37% solids; 3.0 mils DFT
Intermediate	High-Build Epoxy; 77% solids; 5.5 mils DFT
Topcoat	Aliphatic Polyurethane; 61% solids; 1.5 mils DFT

How much abrasive and how much paint do we order? Allow for 20% loss on the coatings and allow for the equivalent of one mil of primer to fill the blast profile. Ignore ladders and other protuberances. The maximum allowable thinning for each paint is 10%.

Solution: When we encounter a problem such as this one, with many parts, the instinctive reaction of most math-adverse people is to throw up our hands. If we are high enough up the company ladder, we delegate the solution to a subordinate or hire a consultant.

The toughest part in solving this type of problem is to get started. Organize the data and break the problem into small segments: area of the tank, the abrasive, the primer, the intermediate coat, and the topcoat. What do we know and what do we need to compute in each segment? If we look at only one part of the solution, we will probably determine that that one part is not too difficult. It is all these parts taken together that may seem overwhelming. We can do it if we proceed *one step at a time*.

Area: We need to determine the area to be blasted and painted. The tank is cylindrical, so we need to find the area of the side wall, A_s, and the area of the roof, A_{roof}. From Chapter 4 on geometry, we find that the area for the side wall is the circumference times the height. Circumference is $2 \cdot \pi \cdot r$ or πd, where the radius, r, is one half the diameter, d. Thus,

$$A_s = 2\pi rh \qquad\qquad A_s = \pi dh$$
$$A_s = 2\pi(25 \text{ ft})(28 \text{ ft}) \quad \text{OR} \quad A_s = \pi(50 \text{ ft})(28 \text{ ft})$$
$$A_s = 4398 \text{ ft}^2 \qquad\qquad A_s = 4398 \text{ ft}^2$$

The area of the side wall is about 4400 ft^2.

The roof is essentially a disc. So to find its area, we can use the formula for the area of a circle.

$$A_{roof} = \pi r^2 \qquad\qquad A_{roof} = \pi\frac{d^2}{4}$$
$$A_{roof} = \pi(25 \text{ ft})^2 \quad \text{OR} \quad A_{roof} = \frac{\pi(50 \text{ ft})^2}{4}$$
$$A_{roof} = 1963 \text{ ft}^2 \qquad\qquad A_{roof} = 1963 \text{ ft}^2$$

The roof area is about 1960 ft^2.

The total external surface area of the tank is the sum of the side area and the roof area: A = 4400 + 1960 = 6360 ft^2.

[We could have used more significant figures, 4398 + 1963 = 6361, but it would still round to 6360.] The total area to be blast cleaned and painted is 6360 ft^2.

Abrasive: We know how much abrasive it will take to blast one square foot (4.0 lb/ft^2), so we multiply this number by the number of square feet. If we do the dimensional analysis, notice how the ft^2 units cancel leaving us with pounds.

$$\frac{4.0 \text{ lb}}{\cancel{\text{ft}^2}} \cdot \frac{1960 \cancel{\text{ft}^2}}{1} = 7480 \text{ lb}$$

This can be rounded to 8000 lb, which is 4 ton. We may actually want to order more than 4 ton to account for waste and to add a safety margin against running out of abrasive. This computation clearly shows that we should not need more than 5 ton.

Primer: The DFT is measured above the profile peaks, so some primer is needed just to fill up the valleys in the profile. We will therefore add one mil to the target DFT before performing the computations. Instead of 3.0 mils of primer we will assume a primer thickness of 4.0 mils. In addition to DFT, we also know percent volume solids ($P_s = 37\% = 0.37$) and the area to be covered ($A = 6360\ ft^2$). We are looking for the volume, V, in gallons. We substitute into the formula that relates our known quantities to our unknown.

$$V = \frac{A \cdot D}{1604 \cdot P_s}$$

$$V = \frac{(6360) \cdot (4.0)}{1604 \cdot (0.37)}$$

$$V = 42.87$$

The theoretical volume is 43 gal. We want to add an additional 20% for wastage. In percent problems, formulate an "is" statement, translate it into a equation, and solve for the unknown. Our "is" statement is:

What is 20% of 42?

$$x = 0.20 \cdot 42$$

After multiplying we find x = 8.6. Thus, we add an additional 9 gal for wastage. The amount of primer needed is 43 + 9 = 52 gal. Paint orders should include some one-gallon kits to avoid splitting kits toward the end of the day. We could order ten 5-gal kits and two 1-gal kits. In addition, if thinning is needed, we would order 10% of 52 gal. The volume of thinner needed (V_T) is 10% of 52 gal, which translates to an equation as:

$$V_T = (0.10) \cdot (52\ gal) = 5.2\ gal$$

We would order 5 or 6 gal of primer thinner. Clean-up solvent is separate, and a 55-gal drum of solvent such as MEK or MIBK would work for all the coatings.

Intermediate Coat: The computation for the volume of intermediate coat parallels that for the primer, except now there is no allowance made for profile because there is none. The known quantities are area $A = 6360\ ft^2$, dry film thickness D = 5.5 mils, and percent volume solids P_s = 77% = 0.77. Substituting these values into the equation above, we can find the volume of intermediate paint needed.

$$V = \frac{A \cdot D}{1604 \cdot P_S}$$

$$V = \frac{(6360) \cdot (5.5)}{1604 \cdot (0.77)}$$

$$V = 28.32$$

The theoretical volume is 28 gal. We must add 20% of this to account for losses. As above, we form an "is" statement.

What is 20% of 28.3?

$$x = 0.20 \cdot 28.3$$

$$x = 5.66$$

This rounds to 5.7 gal. Total volume of intermediate paint needed is 28.3 + 5.7 = 34.0, which we round to 34 gal.

Notice that we kept an extra significant figure during the computations and rounded only at the end. A cautious estimator may round ANY fraction of a gallon up to the next higher gallon. This is the method the post office uses for determining the cost of mailing a package, "$x/lb or fraction thereof."

Therefore, order 34 gal of intermediate paint as six 5-gal kits and four one-gallon kits. The amount of thinner we need is 10% of the 34 gal or 3.4 gal; so order three or four gallons of intermediate thinner. It may be prudent to order 4 gal of intermediate thinner in case we need to thin all 34 gal by the full 10%.

Topcoat: Proceeding as above with the intermediate paint, we identify known quantities as A = 6360 ft^2, D = 1.5 mils, and P_s = 61% = 0.61. Substituting into the formula for volume:

$$V = \frac{A \cdot D}{1604 \cdot P_S}$$

$$V = \frac{(6360) \cdot (1.5)}{1604 \cdot (0.61)}$$

$$V = 9.75$$

Theoretically, 9.8 gal are required. Adding 20% of 9.8 gives (0.20)·(9.8) = 1.96 or 2.0 additional gallons. The total topcoat needed is therefore 9.8 + 2.0 = 11.8, which we round to 12 gal. Thus, order two 5-gallon kits and two one-gallon kits of topcoat. The amount of thinner we need is 10% of the 11 gal, which is (0.10) · (11 gal) = 1.1 gal. So order one or two gallons of topcoat thinner.

Final Order:

Abrasive	8000 lb (4 ton)
Primer	52 gal (ten 5-gal kits; two 1-gal kits)
thinner	6 gal
Intermediate	34 gal (six 5-gal kits; four 1-gal kits)
thinner	4 gal
Topcoat	12 gal (two 5-gal kits; two 1-gal kits)
thinner	2 gal
Clean-up	55-gal drum of MEK

Chapter 8 – Concentration and Other Rates

Objective: To be able to compute VOC content, spreading rate, and pigment volume concentration (PVC) from the paint composition formula and to understand computations involving PVC, soluble salt concentration, and fluid (air/water) flow rates.

8.1 Pigment Volume Concentration (PVC)

SSPC's *Protective Coatings Glossary* defines pigment volume concentration (PVC) as the "ratio of the volume of pigment to the volume of total non-volatile material (i.e., pigment and binder) in a coating. The figure is usually expressed as a percentage." The PVC is a ratio. A ratio compares two numbers by division. The ratio of A *to* B is A divided by B. This division can be written as a fraction, as a decimal, or as a percent.

$$PVC = \frac{\text{volume of pigment}}{\text{volume of total solids}} = \frac{V_p}{V_p + V_b}$$

where

V_p = volume of pigment
V_b = volume of binder

Suppose we start with 5 gal of paint and let the solvent evaporate. If the paint was 80% solids by volume, there will now be 4 gal of dried paint in the bucket. (80% of 5 is 4.) We now separate, theoretically, the pigment from this 4 gal. Suppose we get 1.2 gal of pigment. The remaining 2.8 gal of solids is the binder. We find the PVC by dividing the volume of pigment by the volume of solids:

$$PVC = \frac{1.2\,\cancel{gal}}{4.0\,\cancel{gal}}$$

$$PVC = 0.30 = 30\%$$

Remember, a number can be expressed as a fraction, a decimal, or a percent. PVC is usually expressed as a percent, so this paint would have a PVC of 30%. The PVC can be calculated from the formula, because the formulator knows how much of each ingredient is in the paint. A zinc-rich paint would have a high PVC since zinc dust pigment comprises most of the solids. A high gloss enamel would have a low PVC. At the critical pigment volume concentration, CPVC, there is just enough binder to totally encapsulate the pigment particles. Many important properties are affected at the CPVC.

8.2 Computing Quantities From a Paint Composition Formula

If the formula for a paint is known, we can compute the theoretical value of many different quantities, such as spreading rate, VOC content, PVC, etc. In this section, we will look at a composition specification (SSPC-Paint 25, Type I) and compute various quantities. Table 8.1 is taken from SSPC-Paint 25.

Table 8.1

Composition: SSPC-Paint 25, Type I

INGREDIENTS	REQUIRED		TYPICAL COMPOSITION	
	MIN. WT. %	MAX. WT. %	WT. %	VOL. %
PIGMENT (52.5 ± 2 Wt. %)				
Zinc Oxide	15.0	17.0	8.4	2.2
Iron Oxide (85% minimum Fe_2O_3)	39.0	–	21.5	7.9
Magnesium Silicate	–	33.0	17.3	9.1
Mica	10.0	12.0	5.3	2.7
TOTAL PIGMENT	100		52.5	21.9
VEHICLE (47.5 ± 2 Wt. %)				
Raw Linseed Oil	30.5	–	14.5	23.2
Alkyd Resin Solids	30.5	–	14.5	19.8
Mineral Spirit Thinner	–	39.0	18.5	35.0
Driers	–	–	–	–
TOTAL VEHICLE	100		47.5	78.0
TOTAL			100	100

Selected data from Table 4 of SSPC-Paint 25

Density at min. wt. %	1.48 kg/L	12.3 lb/gal
Flash point at min. wt. %	38°C	100°F
Sag resistance at min. wt. %	0.18 mm	7 mils
Sag resistance at max. wt. %	0.25 mm	10 mils

Let us reconcile the 15-17% by weight of zinc oxide in the "Required" column with the 8.4% by weight in the "Typical Composition" column. Using the midrange value of 16%, zinc oxide must constitute16% by weight *of the pigment*. However, it takes pigment and vehicle to make the paint. Suppose we make a batch of paint weighing 100 lb. Then 52.5 lb of this 100 lb is pigment [52.5% of 100 is 52.5 lb]. Sixteen percent of this 52.5 lb must be zinc oxide. Doing the computation, 16% of 52.5 lb is:

$$(0.16) \cdot (52.5 \text{ lb}) = 8.4 \text{ lb}.$$

Thus, in 100 lb of paint, there will be 8.4 lb of zinc oxide.

The "Typical Composition" column is simply a cookbook recipe for making Paint 25. The quantity of each ingredient is given as a percent in order to be general. As we mentioned above, there will be 8.4 lb of zinc oxide in a 100 lb batch of paint. Similarly, there will be 8.4 kg in a 100 kg batch of paint. Considering this as the recipe for a single batch (100 lb), we could always make a half-batch by dividing all the quantities by 2, or make a double batch by multiplying all the quantities by 2, or multiply by 3 for a triple batch. In fact, we can multiply all the quantities by any number we wish. For example, to make a 250 lb batch, we would multiply each amount by 2.5:

Zinc Oxide	2.5 x 8.4 = 21.0 lb
Iron Oxide	2.5 x 21.5 = 53.75 lb (round to 53.8), etc.

How many gallons would we get if we made a 100 lb batch? We know the weight (100 lb) and we are looking for volume.

Weight (W) and volume (V) are related by the density (D), where:

$$D = \frac{W}{V}$$

or

$$V = \frac{W}{D}$$

or

$$W = VD$$

From Table 4 of Paint 25 we find the density of the paint to be 12.3 lb/gal. Substituting into the formula we find:

$$V = \frac{100 \cancel{lb}}{12.31 \cancel{lb}/gal}$$

OR

$$V = 8.13 \text{ gal}$$

Thus, a 100 lb batch of Paint 25 will yield about 8 gal.

We can use the density concept to measure the ingredients by volume instead of weight. The 100 gal of Paint 25 will have a weight of 1230 lb.

$$(12.3 \text{ lb}/\cancel{gal}) \cdot (100 \cancel{gal}) = 1230 \text{ lb}$$

Zinc oxide will constitute 8.4% of 1230 lb or 103 lb.

$$(0.084) \cdot (1230) = 103.32$$

We must know or look up the density of zinc oxide, 47 lb/gal. Since we also know the weight (103 lb of zinc oxide), we can compute the volume of zinc oxide needed from:

$$V = \frac{103 \cancel{lb}}{47 \cancel{lb}/gal}$$

OR

$$V = 2.19 \text{ gal}$$

So in a 100 gal batch, we need 2.2 gal of zinc oxide. Similar computations can be done with the other ingredients to complete the "volume %" column.

Computing VOC Content: From the data given, let us compute the maximum VOC content of one gallon of Paint 25. The only ingredient that contributes to the VOC content is the mineral spirits thinner, with a maximum value of 39% by weight of the vehicle. Now the vehicle is 47.5% by weight of the total paint. Thus, we weigh one gallon then take 4.75% of this weight and 39% of that number to get the weight of the thinner in one gallon. We work backwards from the weight of one gallon. From the density (Table 4 of Paint 25) we see that one gallon weighs 12.3 lb. The weight of vehicle in one gallon is 47.5% of 12.3 lb or:

$$(0.475) \cdot (12.3 \text{ lb}) = 5.84 \text{ lb}.$$

Thinner comprises 39.0% of this 5.84 lb; the weight of thinner is:

$$(0.390) \cdot (5.84 \text{ lb}) = 2.28 \text{ lb}.$$

This can be rounded to 2.3 lb/gal of VOC.

Computing Spreading Rate: From the data given, we will compute the theoretical spreading rate of a typical composition for Paint 25. From the "Volume %" column we find that the mineral spirits thinner is 35.0% of the volume. This means that 65.0% of the volume is solids. (100% – 35.0% = 65.0%). At a DFT of 2.0 mils, we wish to find the area covered by 1 gal. The formula from Chapter 7 was:

$$\text{Spreading Rate} = \frac{A}{V} = \frac{1604 \cdot P_S}{D}$$

where

A = area in ft^2
V = volume in gal
D = dry film thickness in mils
P_S = percent volume solids
(expressed as a decimal or a fraction)
1604 = the conversion from mil·ft^2 to gal

Substituting for P_s and D we get:

$$\frac{A}{V} = \frac{1604\frac{\cancel{mil} \cdot ft^2}{gal} \cdot (0.650)}{2.0\ \cancel{mils}}$$

Spreading Rate = 521.3 ft^2/gal

So one gallon of Paint 25 will theoretically cover 520 ft^2 at 2.0 mils DFT. Actual spreading rates can be significantly lower.

Appendix D gives theoretical spreading rates for various percent volume solids and dry film thicknesses.

Computing Pigment Volume Concentration (PVC): The pigment volume concentration is the ratio of pigment volume to total volume of solids. For this calculation, we use the data in the "Volume %" column of a typical composition. After subtracting the 35% volume of thinner, we found Paint 25 was 65% solids. From Table 8.1, total pigment is 21.9% of the volume. A simple way to think of this data is that in 100 gal of paint there will be 65 gal of solids of which 21.9 gal is pigment. Forming the pigment to total solids ratio we get:

$$PVC = \frac{\text{volume of pigment}}{\text{volume of total solids}}$$

$$PVC = \frac{21.9\ \cancel{gal}}{65.0\ \cancel{gal}}$$

$$PVC = 0.33692$$

Rounding and then converting to a percent we find the PVC for Paint 25 is 0.337 or 33.7%.

8.3 Salt Concentrations

SSPC Guide 15 (formerly SSPC TU 4) describes procedures for measuring the concentration of soluble salt on the surface. All the procedures follow a common principle: a known quantity of salt-free water comes in contact with a known area of contaminated surface and some of the salt dissolves in the water. Depending of the surface, the nature of the salts, and the method used to extract the salts, typically, only from 15 to 85% of the salts on the surface are dissolved in the water. Currently, the efficiency of various extraction methods is not well established.

We measure the amount of dissolved salt and compute the amount of salt on the surface in micrograms per square centimeter. Before doing a sample analysis, we will discuss some chemistry terms. Chlorides, sulfates, and nitrates are commonly found ions that are very corrosive to steel; however, for simplicity in understanding the calculations, chlorides will be used to represent calculations involving salts.

Surface Concentration: The most common damaging ion to painted steel is the chloride ion. This can be carried to the steel surface as sodium chloride, calcium chloride, ammonium chloride, and others. In preparing for painting, we are concerned with only the chloride ion and do not care about the other part of the compound. Surface concentration is the weight of chloride ion per unit surface area, and the common units are micrograms per square centimeter, $\mu g/cm^2$. If 30 μg of chloride are spread over 5 cm^2, the surface concentration will be:

$$(30\ \mu g) \div (5\ cm^2) = 6\ \mu g/cm^2.$$

Volume Concentration (ppm): When water is placed in contact with a contaminated surface, some chloride ions will dissolve in the water. The water containing the dissolved ions is called the extract. The volume of extract is measured in milliliters, mL. There are 1000 mL in one liter, which is slightly more than one quart, and 30 mL in one fluid ounce. The weight of dissolved chloride is measured in grams. There are 454 grams in a pound. The weight of dissolved chloride, W, is divided by the volume, V, of the extract to get the concentration, C.

$$C = \frac{W}{V}$$

Example: If 210 μg of chloride are dissolved in 5.0 mL of extract, find the concentration in g/mL.

Solution: We know the weight and the volume and we want to find concentration, so we substitute into the formula and then convert from grams to micrograms.

$$C = \frac{210\ \mu g}{5.0\ mL}$$

$$C = \frac{210\ \cancel{\mu g}}{5.0\ mL} \cdot \frac{1\ g}{1{,}000{,}000\ \cancel{\mu g}}$$

$$C = 0.000042 \frac{g}{mL}$$

The chloride concentration of the extract is 0.000042 g/mL. This can also be expressed as 0.042 mg/mL or 42 μg/mL.

Instead of measuring concentration as weight per unit *volume,* we can compare the weight of chloride to the *weight* of solution. Dilute aqueous (water) solutions have a density of approximately 1.0 g/mL. Thus, the volume of the solution in milliliters is numerically equal to the weight of the solution in grams. A volume of 15 mL weighs 15 g, 5 mL weighs 5 g, etc. Defining concentration as weight of dissolved salt divided by weight of solution, the concentration is a dimensionless quantity, i.e., the units cancel.

$$\text{concentration} = \frac{\text{weight of dissolved salt}}{\text{weight of extract}}$$

Example: If 210 μg of chloride are dissolved in 5.0 mL of extract, find the concentration by comparing the weight of salt to the weight of extract.
Solution: The 5.0 mL of extract has a weight of 5.0 g. Substituting in the definition for concentration and multiplying by the conversion factor from micrograms to grams:

$$C = \frac{210\ \cancel{\mu g}}{5\ g} \cdot \frac{1\ g}{1{,}000{,}000\ \cancel{\mu g}}$$

$$C = 0.000042$$

OR

$$C = \frac{42}{1{,}000{,}000}$$

The concentration is 42 parts chloride compared to one million parts

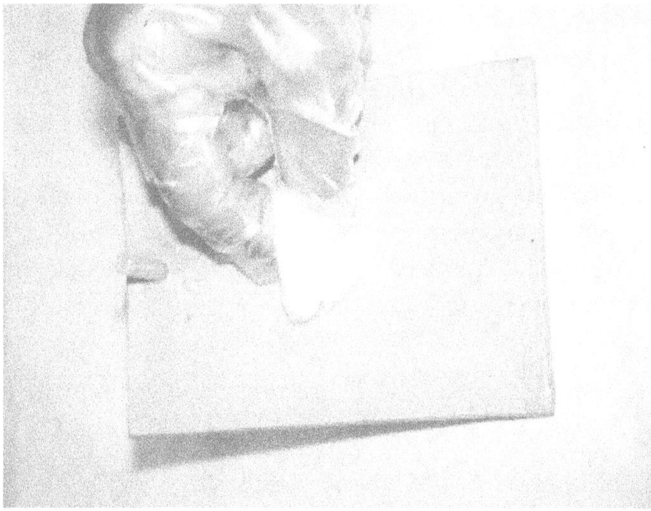

Figure 1. Swabbing with cotton ball wetted with deionized water to extract salts from a steel surface. Swabbing is one of several methods of salt extraction.

solution. In this form, concentration is expressed as 42 parts per million or 42 ppm. If the appropriate conversion factors are used, dividing the weight of dissolved salt in micrograms by the volume of solution in milliliters gives the concentration in parts per million.

Formula for Concentration in ppm

$$\text{concentration (in ppm)} = \frac{\text{weight of dissolved salt (in } \mu\text{g)}}{\text{volume of solution (in mL)}}$$

assuming a dilute aqueous solution of density 1.0 g/mL.

Converting from Volume Concentration to Surface Concentration: When a contaminated surface is washed or otherwise comes in contact with water, some of the salts dissolve in the water. We measure the weight of salt (more specifically chloride ions) dissolved in the water. If we also know the area of the surface from which the salt was extracted, we can

compute the surface chloride concentration in grams per square centimeter.
Example: Suppose an area 6 x 6 inches is marked off. Cotton balls dipped in deionized water are used to wash the surface within the marked square. Assume that the extraction efficiency of the method is 35%. After performing the procedure, we are left with 25.0 mL of extract containing the dissolved chloride from the surface. Chemical tests determine the chloride concentration of the extract to be 38 ppm. What is the surface concentration of chloride in micrograms per square centimeter?
Solution: We should work in a single system of units. These problems are usually easier in metric units, so our first step is to find the area in square centimeters.

$$A = (6 \text{ in}) \cdot (6 \text{ in})$$

$$A = 36 \cancel{\text{in}^2} \cdot \frac{6.45 \text{ cm}^2}{1 \cancel{\text{in}^2}}$$

$$A = 232 \text{ cm}^2$$

The washed area is 232 cm^2. We can carry extra significant figures here.

We know the area, so we must find the total weight of chloride on that area in order to get the surface concentration. We assumed that 35% of the total chloride is dissolved in the extract. So if we can find the total weight of chloride in the extract, we can determine the weight of chloride originally on the washed area. The next step is to find the weight of chloride in the extract.

The concentration of the extract was measured to be 38 ppm, which is equivalent to 38 μg/mL. The weight of chloride is then:

$$(38 \ \mu\text{g/mL}) \cdot (25.0 \text{ mL}) = 950 \ \mu\text{g}$$

So 950 μg of chloride from the surface was in the extract. This represents only 35% of the total chloride on the surface. Forming an "is" statement and then translating to an equation:

$$950 \ \mu\text{g is } 35\% \text{ of what?}$$

$$950 = 0.35 \cdot x$$

Dividing both sides of the equation by 0.35 and reversing the equation we get:

$$x = \frac{950\ \mu g}{0.35}$$

$$x = 2714\ \mu g$$

This can be rounded to 2710 μg of chloride on the surface.

Surface concentration is total weight of chloride divided by total surface area.

$$C_{surface} = \frac{2710\ \mu g}{232\ cm^2}$$

$$C_{surface} = 11.68 \frac{\mu g}{cm^2}$$

From our data we conclude that the chloride surface concentration is 12 $\mu g/cm^2$.

Putting the ideas of the previous example together, we find that the solution concentration can be converted to a surface concentration:

Formula for Surface Concentration

$$\text{surface concentration } (\mu g/cm^2) = \frac{\text{extract concentration (in ppm)} \cdot \text{volume of extract (in mL)}}{\text{area of surface extracted (in cm}^2\text{)}}$$

Substituting numbers directly into the formula we get

$$C_{surface} = \frac{(38\ ppm) \cdot (25.0\ mL)}{232\ cm^2}$$

And since 1 ppm = 1 $\mu g/mL$, we get

$$C_{surface} = \frac{\left(38 \frac{\mu g}{\cancel{mL}}\right) \cdot (25.0 \cancel{mL})}{232\ cm^2}$$

$$C_{surface} = 4.09 \frac{\mu g}{cm^2}$$

But this is only 35% of the total chloride on the surface. We form an "is" statement:

$$4.09 \text{ is } 35\% \text{ of what?}$$

$$4.09 = 0.35 \cdot x$$

which we solve for x to get 11.7. The surface concentration is 12 $\mu g/cm^2$, the same as we found above.

8.4 Volume Flow Rate

The rate of air flow through a containment or through a detector is often specified as so many cubic feet per minute (CFM) or cubic meters per minute. This is specifying volume per unit time. This flow can be described with any volume unit divided by any time unit. Liter is a measure of volume and day is a measure of time, so rate of flow could be measured in liters per day.

If the cross-sectional area is constant, such as in a pipe, the rate of material flow through the pipe is the cross-sectional area times the velocity of the material. Here we assume laminar flow, that is, smooth flow without turbulence. We also ignore frictional losses so that the fluid near the sides of the containment is moving at the same speed as fluid in the center.

Flow Rate Formula

$$\text{flow rate} = (\text{cross sectional area}) \cdot (\text{velocity of material})$$

Or using letters,

$$Q = A \cdot v$$

where

Q = flow rate
A = cross sectional area
v = velocity of material

Example: Water is moving through a 6-inch diameter pipe at 10 ft/s. What is the flow rate in cubic feet per minute and in gallons per minute?
Solution: We begin by identifying the known quantities and assigning letters to them. The pipe diameter, d = 6.0 inches; the velocity of the water, v = 10 ft/s. We are looking for flow rate, Q. We could use the formula

above if we knew the area, A. But we can find area because we know the diameter. The radius, r, of the pipe is half the diameter, so r = 3.0 inches. Solving for A we get:

$$A = \pi \cdot r^2 \qquad\qquad A = \pi d^2/4$$
$$A = \pi \cdot (3.0\text{ in})^2 \qquad \text{OR} \qquad A = \pi(6.0\text{ in})^2/4$$
$$A = 28.27\text{ in}^2 \qquad\qquad A = 28.27\text{ in}^2$$

Substituting into the formula for flow rate and multiplying by the conversion factor for in^2 and ft^2:

$$Q = (28.27\,\cancel{\text{in}^2}) \cdot \left(10\frac{\text{ft}}{\text{s}}\right) \cdot \frac{1\text{ ft}^2}{144\,\cancel{\text{in}^2}}$$

$$Q = 1.963\frac{\text{ft}^3}{\text{s}}$$

So the water is flowing at about 2.0 ft^3/s through the pipe.

To convert to other units we must multiply by the appropriate conversion factors.

$$Q = \frac{1.96\text{ ft}^3}{\text{s}} \cdot \frac{60\text{ s}}{1\text{ min}}$$

$$Q = 118\frac{\text{ft}^3}{\text{min}}$$

OR

$$Q = \frac{1.96\,\cancel{\text{ft}^3}}{\cancel{\text{s}}} \cdot \frac{60\,\cancel{\text{s}}}{1\text{ min}} \cdot \frac{7.48\text{ gal}}{1\,\cancel{\text{ft}^3}}$$

$$Q = 879.6\frac{\text{gal}}{\text{min}}$$

Rounding, we find the flow rate to be 120 ft^3/min or 880 gal/min.

In this book, the formulas assume laminar flow, which is flow without turbulence. This assumption becomes less valid as the velocity of the fluid increases. Losses due to friction with the side wall and turbulence can be accounted for by introducing a coefficient that is determined empirically. For example, to compute the flow through a waterjet nozzle, the effective area is the order of 0.7 times the actual cross-sectional area of the nozzle orifice.

Example: What dust collector capacity (CFM) is required to keep a cross draft of 100 ft/min through a rectangular containment 8 ft wide, 10 ft high, and 60 ft long? Ignore leaks and frictional losses.

Figure 2. Containment tarps for an overhead railway.

Solution: We know the velocity of the air through the containment and we can find the cross-sectional area. The theoretical collector capacity will then be the rate of flow.

Since the cross-section is a rectangle, we use the formula, area equals length times width, or in this case, height times width:

$$A = h \cdot w$$
$$A = (10 \text{ ft}) \cdot (8 \text{ ft})$$
$$A = 80 \text{ ft}^2$$

Substituting into the flow rate formula:

$$Q = A \cdot v$$
$$Q = (80 \text{ ft}^2) \cdot (100 \text{ ft/min})$$
$$Q = 8000 \text{ ft}^3/\text{min}$$

The theoretical fan capacity must be 8000 ft^3/min or 8000 CFM. In practice, a higher rated fan would be required to account for losses. Notice that the length of the containment, 60 ft, was not used in the computation.

Blast cleaned steel in a tank or other enclosed space can be prevented from rusting if the relative humidity is kept low enough. While blast cleaning, a constant airflow is required to control dust. This is usually stated as so many air changes per hour. The air must be dehumidified before entering the tank. DH equipment contractors have guides that give the capabilities of their equipment, usually expressed as cubic feet per minute or cubic meters per hour. The problem is to calculate the correct size of equipment for the job.

Example: A tank 90 ft in diameter with a roof 6.0 ft high is having the blast protected with dehumidification. Find the capacity of the dehumidification equipment (in ft^3/min) if two air changes are required each hour.
Solution: The capacity is a rate measured in units of volume per unit time. We must compute the volume of air treated in a given time interval and form the fraction with volume in the numerator and time in the denominator. We then convert to the desired units.

Since there are two air changes required each hour, the volume of air treated in one hour will be twice the volume of the tank. We first compute the tank volume from the formula for the volume of a cylinder:

$$V = \pi r^2 h \quad \text{OR} \quad V = \pi \frac{d^2}{4} h$$

We know the diameter, d = 90 ft, and the height, h = 6.0 ft, so we use the second formula. Substituting into the second formula we get:

$$V = \pi\frac{(90\ \text{ft})^2}{4}(6.0\ \text{ft})$$

$$V = 38{,}170\ \text{ft}^3$$

which we round to 38,200 ft^3. (We can carry an extra significant figure in the intermediate calculations.) The total volume of air treated each hour is then twice this amount or 76,400 ft^3. Rate of air flow is then 76,400 ft^3/h. The next step is to convert to ft^3/min.

$$\frac{76{,}400\ \text{ft}^3}{1\ \cancel{\text{hr}}}\cdot\frac{1\ \cancel{\text{hr}}}{60\ \text{min}} = 1273\frac{\text{ft}^3}{\text{min}}$$

Rounding, we determine that the DH equipment must have a capacity of at least 1300 ft^3/min or 1300 CFM.

This type of calculation is discussed in *SSPC-TR 3/NACE 6A192 (2000 Revision): Dehumidification and Temperature Control During Surface Preparation, Application, and Curing for Coatings/ Linings of Steel Tanks, Vessels, and Other Enclosed Areas.*

Instead of breaking the example problem into pieces, we could have computed the volume of the tank (38,200 ft^3) and then substituted into the formula. We are given the

Formula to Compute the Flow Capacity of a Dehumidifier (U.S. Customary Version)

$$X = \frac{(V_i)(RAC)}{60}$$

where

X = air-flow capacity in ft^3/min

V_i = the internal volume of the space minus the volume of any obstructions in ft^3

RAC = the required air changes per hour

60 = the conversion factor from hours to minutes

Formula to Compute the Flow Capacity of a Dehumidifier (Metric Version)

$$X = (V_i)(RAC)$$

where

X = air-flow capacity in m^3/h

V_i = the internal volume of the space minus the volume of any obstructions in m^3

RAC = the required air changes per hour

number of air changes per hour, RAC = 2. Then:

$$X = \frac{(38{,}200)(2)}{60}$$

$$X = 1273$$

Rounding, the needed capacity of the DH equipment is 1300 ft^3/min, which is the same as we got before.

If metric quantities are specified, the DH equipment capacity is computed in cubic meters per hour. The conversion factor of 60 is not needed.

Figure 3. DH equipment installed beside a shrouded tank.
Courtesy Enviro-Air Control Corp.

Example: Find the capacity of dehumidification equipment that can dehumidify a 12-m-diameter, 5-m-high enclosed space if 3 air changes are required each hour.

Solution: We know diameter and height, so we must first calculate the volume before we can substitute into the formula.

$$V = \pi\frac{d^2}{4}h$$

$$V = \pi\frac{(12\ m)^2}{4}(5\ m)$$

$$V = 565.49\ m^3$$

Rounding to V = 565 m^3 and substituting into the capacity formula we get:

$$X = (V_i) \cdot (RAC)$$

$$X = (565\ m^3) \cdot (3/h)$$

$$X = 1695\ m^3/h$$

Thus, we need DH equipment rated at least 1700 m^3/h.

Chapter 9 – Making Judgments from Measurements and Computations

Objective: To develop the ability to make sound judgments from measurements and computations. To be able to ascertain compliance or noncompliance with specified values for dry-film thickness, relative humidity, and dew point.

9.1 Accuracy and Precision

Standards and specifications include target values that must be met. Some of these are direct measurements and some are calculated from other measurements. Any measurement contains some uncertainty, even if state-of-the-art instruments are used.

Accuracy and precision are easily confused concepts. An accurate measurement agrees with the established standard. With a perfect instrument under ideal conditions, our answer will be accurate. A measurement is precise if we get the same value over and over. Precision refers to repeatability or reproducibility of the measurement.

Suppose we have a sphere with a diameter, compared to national standards, of 10.00 inches. We measure the diameter three times with a cheap ruler and get 9.8, 10.0, and 10.2 inches as our three measurements. The average of these three measurements is 10.0 inches. This is a very accurate measurement. However, the method was only precise to 2 parts in 100 or 2%. A lab technician uses a caliper to measure the diameter of the sphere and records measurements of 9.87, 9.86, and 9.88 inches. The average value is 9.87 inches. The caliper measurement is not as accurate as the ruler because it is not as close to the accepted (true) value of 10.00 inches. The caliper is much more precise than the ruler. The range of caliper measurements is 0.02 inches while the range with the ruler is 0.4 inches. The caliper is twenty times more precise (0.02 x 20 = 0.4).

With the caliper, every measurement was less than the true value. This is evidence of a systematic error. Systematic errors are always off in the same direction. In this example, the caliper was not properly adjusted.

A random error is just as likely to cause a measurement to be too large or too small. Why did the caliper not read exactly the same for all three measurements? Perhaps the sphere was not perfectly round. We are just as likely to measure a high spot as a low spot. In that case, variation in measurement was caused by the object being measured. Fluctuations in the measurements can also be caused by the technique itself; e.g., if we had to visually estimate the distance between the smallest markings on the caliper.

Systematic errors can often be corrected, such as when the base metal reading is subtracted from the reading of a Type I pull-off DFT gauge. Random errors are addressed by taking an average of several readings. If all readings are close to each other, fewer readings are necessary to obtain a meaningful average value. If there are wide fluctuations among the readings, then a large number of readings should be taken when computing the average. The more readings we take, the higher the probability that the average will be close to the "true" value.

ASTM International has several documents that explain how to determine accuracy and precision, such as *ASTM E 691: Standard Practice for Conducting an Interlaboratory Study to Determine the Precision of a Test Method.* Most ASTM test methods have a precision statement such as this one from *ASTM D 1475: Standard Test Method for Density of Liquid Coatings, Inks, and Related Products,* "Two results, each the mean of duplicate determinations, obtained by operators in different laboratories should be considered suspect if they differ by more than 1.8% relative." This is at the 95% confidence level (see Chapter 6).

Example: We measure the density of HydeRite paint in our lab in accordance with ASTM D 1475 and find that the average of three runs is 16.4 lb/gal. A sample of paint from the same batch had been sent to U-Test-It Labs for measurement of density. Within what range of values can the U-Test-It measurement be considered in agreement with our value of 16.4 lb/gal?
Solution: Since ASTM D 1475 allows for a 1.8% deviation, we must first compute 1.8% of 16.4. Computing 1.8% of 16.4 we get $(0.018) \cdot (16.4) = 0.2952$ or about 0.3 lb/gal. The lowest density U-Test-It could measure without being suspect would be 16.4 – 0.3 or 16.1 lb/gal. Similarly, the highest value would be 16.4 + 0.3 or 16.7 lb/gal. Thus, as long as the density measurement of HydeRite paint at U-Test-It lab falls between 16.1 and 16.7 lb/gal, their measurements can be considered to be in agreement with ours (with 95% confidence).

9.2 DFT per SSPC-PA 2

If two inspectors each measure the DFT of paint on a steel plate, they will most likely get two different numbers, but their measurements will probably be close to each other. If one of the inspectors measures the plate again, the second DFT measurement will most likely not be exactly the same as the first measurement. What if the inspector uses a different gauge? Which measurement do we use to determine if the spec has been met?

In SSPC-PA 2, the average of three gauge readings constitutes a spot measurement. If the gauge is accurate to the nearest 0.1 mil, then the average should not be stated more accurately than the nearest 0.1 mil.

Example: The three gauge readings are 2.4, 2.6, and 2.9 mils. What is the spot measurement?
Solution: To find the average value, add the three readings and divide the sum by 3:

$$(2.4 + 2.6 + 2.9) \div 3 = 7.9 \div 3 = 2.633$$

We round to the nearest 0.1 mil and record the spot measurement as 2.6 mils.

Example: Three gauge readings are 5.7, 6.1, and 6.1 mils. What is the spot measurement?
Solution: Find the average of the gauge readings:

$$(5.7 + 6.1 + 6.1) \div 3 = 5.9667$$

We round to the nearest 0.1 mil and record the spot measurement as 6.0 mils. Here, the zero is significant, so we should not record the measurement as 6 mils. The number of significant figures reflects the precision of the instrument.

Example: Three gauge readings are 28, 34, and 29 mils. What is the spot measurement?
Solution: Find the average of the gauge readings:

$$(28 + 34 + 29) \div 3 = 91 \div 3 = 30.33$$

We round to the nearest 1 mil and record the spot measurement as 30 mils. In this instance, we know from the context of the problem that the zero after the 3 is significant. So 30 has two significant figures.

Example: Three gauge readings are 1.03, 0.96, and 1.12 mils. What is the spot measurement?
Solution: Find the average of the gauge readings:

$$(1.03 + 0.96 + 1.12) \div 3 = 1.03667$$

We round to the nearest 0.01 mil and record the spot measurement as 1.04 mils.

SSPC-PA 2 stipulates that a spot measurement (the average of three gauge readings) should not be less than 80% of the specified minimum DFT nor more than 120% of the specified maximum DFT. A practical procedure before starting the inspection is to compute the upper and the lower limits for a spot measurement. Once this allowable window is determined, we will immediately know if the spot conforms to the specification.

Example: The DFT specification for a zinc-rich primer calls for a minimum of 3.0 mils and a maximum of 8.0 mils. What are the minimum and the maximum allowable spot measurements?
Solution: PA 2 requires that the average of the spot measurements must fall between 3.0 and 8.0 mils. However, it is permissible for an individual spot measurement to fall outside this range, provided it is not too far outside the range. We first find the lower limit by computing 80% of 3.0 mils. Forming an "is" statement and translating it into an equation we get:

What	is	80%	of	3.0?
x	=	0.80	·	3.0

After multiplying, we get 2.4 mils as the minimum allowable spot measurement.

Repeat the procedure for the upper limit, i.e., the maximum allowable spot measurement.

What	is	120%	of	8.0?
x	=	1.20	·	8.0

After multiplying we get 9.6 mils as the maximum allowable spot measurement.

As we inspect the structure, we can immediately identify a spot measurement less than 2.4 mils or greater than 9.6 mils as not being in compliance with the specification. All spot measurements between these two values are allowed. If we encounter a non-conforming spot, we can then take additional measurements to determine the non-conforming area. Even though an individual spot measurement may be above or below the specified limits, the average of the spot measurements must be between 3.0 and 8.0 mils.

According to PA 2, the DFT of the coating on a 100 ft^2 area is the average of five spot measurements made at random locations. The average should be stated with the same number of decimal places as the spot measurements.

Example: The five spot measurements are 7.3, 6.6, 6.9, 7.1, and 6.8. The specified minimum DFT was 7.0 mils. Has the specification been met?
Solution: The DFT is the average of the five spot measurements:

$$(7.3 + 6.6 + 6.9 + 7.1 + 6.8) \div 5 = 6.94$$

We round to the nearest 0.1 mil and get a DFT of 6.9 mils. According to these measurements the DFT has not been met. It is shy by 0.1 mil.

If the minimum DFT had been specified as 7 mils, then we could round the 6.9 to 7, and the DFT specification would be met. The difference between 7 mils and 7.0 mils is one of precision. Specifying 7 mils means closer to 7 than to 6 mils. Specifying 7.0 mils means closer to 7.0 than to 6.9 mils. The DFT should not be specified to a precision exceeding that of the available gauge. We should not specify DFT of 7.250 mils if the gauge is only precise to the nearest 0.1 mil or 0.01 mil. In the case of DFT, the gauge may not be the limiting factor, but rather the blast profile. Even with a gauge that can read to the nearest 0.001 mil, the surface irregularity may limit the precision of the DFT measurement to the nearest 0.1 mil or even the nearest 1 mil.

The Appendix of SSPC-PA 2 contains a numerical example where an individual spot is less than the specified DFT but the average of the five spots meets the specification.

9.3 Quality Assurance Acceptance Criteria

Many ASTM test methods contain precision and bias statements. Specification developers should be cognizant of the fact that these statements exist. Quality assurance test criteria should not be more stringent than the interlaboratory precision of the test methods used.

A common quality assurance test is density (weight per gallon) as measured by ASTM D 1475. The interlaboratory reproducibility at the 95% confidence level tells us that because of errors associated with the method, any two measurements of the same material can differ by as much as 1.8%, without the measurements being considered suspect. This only represents the precision of the measurement technique and does not account for normal variances in the manufactured product.

The acceptable range for paint density must be stated. For example, a composition specification may state this requirement as 10.0 ± 0.2 lb/gal, 10.0 lb/gal ± 2%, or as a range from 9.8 to 10.2 lb/gal. The manufacturer of proprietary products should provide this information.

Example: The tech data sheet lists the density of LustreLux paint to be between 9.8 and 10.2 lb/gal. As the owner's representative, what criteria should we use to determine whether or not a shipment of LustreLux meets the density specification? In other words, what is the acceptable range of density measured by us?

Solution: We need to compute the high and the low limits. If the manufacturer's lab measures the density to be 9.8 lb/gal, the product meets the low end of the specification and the paint is shipped to the job. Because of the lab-to-lab variation of 1.8%, our lab may measure the density of this sample to be as low as 9.8 less 1.8% of 9.8 lb/gal:

$$9.8 - (0.018) \cdot (9.8) = 9.6 \text{ or } 9.6 \text{ lb/gal.}$$

Similarly for the high end, the manufacturer may measure density to be 10.2 lb/gal, which meets the high end of the specification, and ship the paint to the job. Because of the lab-to-lab variation of 1.8%, our lab may measure the density of this sample to be as high as 10.2 plus 1.8% of 10.2:

$$10.2 + (0.018) \cdot (10.2) = 10.4 \text{ or } 10.4 \text{ lb/gal.}$$

Our pass/fail criteria to accept a batch of paint should therefore be 9.6 to 10.4 lb/gal.

Not all ASTM standards have precision and bias statements. How do we establish limits of acceptance? Where precision and bias data are not available, one method is to determine the reproducibility of the measurement method in our own lab. For a given test method, determine the mean and the standard deviation of a minimum of five measurements taken on the originally tested and certified material. To pass the quality assurance test, the measurement of the test sample should fall within two standard deviations of the target value (see Chapter 6). The contracting parties must agree on a target value. Two standard deviations represents the 95% confidence level.

9.4 Dew Point and Relative Humidity

Good painting practice dictates that the surface temperature of steel should be at least 5°F (3°C) above the dew point during coating application in order to ensure moisture has not condensed on the surface. Some zinc-rich coatings and moisture-cure polyurethane coatings require a minimum level of humidity in the air for curing. Other coatings cure more slowly if the relative humidity is high. Dehumidification systems may be used to lower the dew point to retard rusting of a blast-cleaned steel surface. This section describes a procedure for computing dew point and relative humidity from temperature measurements.

Air is composed of independent molecules of nitrogen, oxygen, carbon dioxide, other elements and compounds, and water. Think of each H_2O "water" molecule by itself bouncing around with all the other molecules that make up the air. Temperature is a measure of how fast all of the molecules are moving. If the average speed of the gas molecules increases, the temperature will rise. If all the gas molecules were to stop, the temperature would be absolute zero, –459°F (–273°C). The temperature of a substance with no molecular motion is zero kelvin, 0 K, on the absolute temperature scale. Substances will not be a gas near absolute zero because they will liquefy or solidify before getting that cold.

As the temperature falls, the molecules slow down enough so that they stick together. Water molecules are attracted to other water molecules. At high enough temperatures, the molecules keep bouncing into each other with enough force to break them apart as fast as they clump together. If there are enough water molecules in the air that are moving slowly enough to form large clumps, the water will precipitate out as dew (or frost if it is very cold). The temperature at which dew forms is called the **dew point**.

The number of water molecules in the air is a measure of the humidity. There is absolute humidity and relative humidity. **Absolute humidity** is a measure of the number of water molecules in a given volume of air. This is usually measured as weight of water per unit volume (grains per cubic foot or grams per cubic meter). **Relative humidity** (RH) is the ratio of the amount of water actually in the air to the maximum amount of water the air could hold at that temperature and pressure. Since RH is a ratio, the units cancel; so RH is a dimensionless number. RH is usually expressed as a percent. A relative humidity of 80% means that the number of water molecules in the air is 80% of the maximum number the air could hold.

$$\text{relative humidity} = \frac{\text{actual weight of water in a unit volume of air (gr/ft}^3\text{)}}{\text{maximum weight of water a unit volume of air can hold (gr/ft}^3\text{)}}$$

When air is saturated, it can hold no more water, and the relative humidity is 100%. Relative humidity is often reported on the weather news.

Relative humidity depends on temperature and pressure. In cold air, the molecules are moving more slowly so it is easier for the water molecules to stick together and form drops, i.e., condensation. Hence, cold air cannot hold as much water as warm air, where the molecules are moving faster. As the temperature drops, the maximum amount of water the air can hold decreases while the actual amount of water in the air remains fixed. If the temperature drops low enough, the air becomes saturated, it cannot hold any more water, and the RH is 100%. A further drop in temperature will cause the water to condense.

Grain is a unit of weight in the U.S. customary system and the abbreviation is gr. Historically, a grain was the weight of a grain of wheat. Weight of water per unit volume can be measured as grains per cubic foot (gr/ft^3). In the metric system, weight or mass is measured in kilograms, grams, or milligrams, with the abbreviations kg, g, and mg, respectively. Absolute humidity can be measured in units of mass per unit volume, such as kilograms per cubic meter (kg/m^3) or milligrams per liter (mg/L). We must be careful not to confuse grains (gr) with grams (g). One grain is about 65 milligrams.

Air near a steel surface, e.g., in the bottom of the pits of the profile, will assume the temperature of the steel. Under certain conditions, the temperature in the pit may be slightly lower than the air temperature and water may condense out of the air and onto the steel. Good painting practice dictates that the surface temperature be 5°F (3°C) above the dew point before paint application. The 5°F (3°C) is a safety margin to avoid moisture condensation.

How do we measure dew point and relative humidity? The device most commonly used is the sling psychrometer (see **Figure 1**). Hand-held electronic gauges have been developed that will measure surface temperature, air temperature, and relative humidity. A sling psychrometer consists of two thermometers that are twirled so that the bulb of each thermometer is moving rapidly through the air. The bulb of one of the thermometers is covered by a wet muslin cloth.

Evaporation is a cooling process. When water evaporates, it absorbs heat from whatever it is in contact with. When we step out of a swimming pool, we feel cool because the evaporating water absorbs heat from our body. The same principle applies to the wet bulb thermometer. As the water in the muslin evaporates, it absorbs heat from the bulb, thereby cooling the mercury in the thermometer. The wet bulb thermometer will give a lower reading than the dry bulb thermometer. This temperature difference is called the temperature depression and is used to find dew point and RH.

Figure 1. Sling psychrometer.
Courtesy Elcometer, Inc.

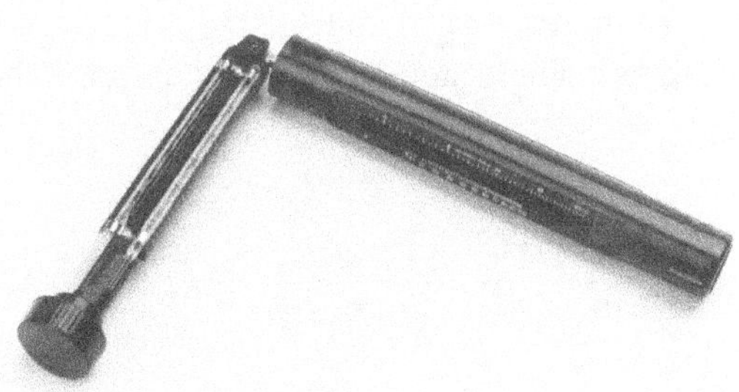

If the RH is low, water will evaporate quickly from the wet bulb resulting in a large temperature depression. If the air is saturated, no water will evaporate and the two thermometers will read the same temperature, i.e., the temperature depression will be zero.

Psychrometric tables relating temperature depression to dew point and relative humidity were first published by the U.S. government in 1900 and are now in the public domain. Copies can be obtained from suppliers of inspection equipment or on the Web at http://www.srh.noaa.gov/elp/psych/psych.html. The formulas for computing these quantities are quite involved, so the practical solution is to resort to tables. Some of these tables are reproduced in Appendix E.

Air pressure also affects dew point and relative humidity, so tables have been constructed for various atmospheric pressures. The tables (created in 1900) give pressures such as "30.0 inches," which means the atmospheric pressure will support a 30.0-inch high column of mercury in a barometer. If we convert inches to millimeters we get:

$$30.0 \text{ inches} \cdot \frac{25.4 \text{ mm}}{1 \text{ inch}} = 762 \text{ mm}$$

The mercury column in a barometer would be 762 mm high. The modern metric unit for pressure is the pascal (Pa). The chemical symbol for mercury is Hg. One millimeter of mercury is equivalent to 133.32 pascals. We can convert 762 mm of mercury to pascals:

$$762 \cancel{\text{mm Hg}} \cdot \frac{133.32 \text{ Pa}}{1 \cancel{\text{mm Hg}}} = 101{,}600 \text{ Pa}$$

OR

101.6 kPa

Thus, 30.0 inches of mercury is equal to 101.6 kPa. These two conversions can be combined into a single step so we can convert inches of mercury directly to kilopascals:

$$1 \cancel{\text{in Hg}} \cdot \frac{25.40 \cancel{\text{mm Hg}}}{1 \cancel{\text{in Hg}}} \cdot \frac{133.32 \cancel{\text{Pa}}}{1 \cancel{\text{mm Hg}}} \cdot \frac{1 \text{ kPa}}{1000 \cancel{\text{Pa}}} = 3.386 \text{ kPa}$$

Therefore, to four significant figures, a pressure of one inch of mercury is equal to 3.386 kPa.

Example: At the job in Pittsburgh, the air pressure is 29.8 inches of mercury. The air temperature, t, is 78°F. The wet bulb temperature, t′, is 72°F. The surface temperature of the steel, t_s, is 81°F.

a) What is the dew point?
b) What is the relative humidity?
c) Are conditions suitable for applying paint?

Solution: a) The first step is to compute the temperature depression of the wet bulb:

$$t - t' = 78° - 72° = 6°$$

Since the tables are not constructed for tenths of an inch of pressure, we will use the table for 30.0 inches of mercury. Matching the row for 78°F with the column for 6° depression we find the dew point temperature to be 69°F.

b) Referring to the relative humidity table for 30.0 inches of mercury, we locate the intersection of the row for 78° and the column for 6° depression and find RH to be 75%.

c) The difference between the surface temperature and the dew point temperature is 81° – 69° = 12°, which is greater than or equal to 5°F, so conditions are suitable for paint application.

Chapter 10 – Cost Computations

Objective: To learn methods of computing the future value of an investment, or estimating the cost of future maintenance. To become more competent in comparing the costs of two competing coating systems.

10.1 The Time Value of Money

This chapter discusses methods of computing the future value of an investment and the amount that must be invested now to have a given amount at a particular future time if we know the annual rate of growth. Compound interest and continuous growth are compared.

When comparing the economics of two coating systems, we can compare the cost per year in today's dollars over a given lifetime, or we can compare the amount of money that we must spend and invest now to cover initial painting plus future maintenance costs. In the case of planned maintenance, we must assume an annual inflation rate as well as the rate of return on our investments. Both of these assumptions are guesses, so precise computations are not usually warranted.

Many business, accounting, and math books derive the compound interest formula:

$$A = P\left(1 + \frac{r}{n}\right)^{nt}$$

where

A = amount after t years
P = principal (amount at time zero)
r = annual rate expressed as a decimal
n = number of compounding periods per year
(n = 4 for quarterly compounding)
t = time in years
r/n = rate per compounding period
nt = total number of compounding periods in t years

If compounding is done only once per year, n = 1, the formula becomes:

$$A = P(1 + r)^t$$

If compounding is done continuously (n approaches infinity), the formula reduces to:

$$A = P e^{rt}$$

where $e \approx 2.71828$ is the base of the natural logarithms. The theory of logarithms is discussed in calculus textbooks and is beyond the scope of this book.

Since predicting the future (inflation rate, labor and material costs, etc.) is guess work anyway, any of these formulas will give us a ball park figure. The last one is the simplest to use if we have a scientific calculator.

The other formulas also require a scientific calculator, but tables often can be found for the factor $(1 + r)^t$. We will do examples using both formulas.

Example: If \$5000 is invested now at 6%, what will be the amount after 8 years if compounding is done (a) annually? (b) monthly? (c) continuously?

Solution (a): We assign letters to our known quantities. P = \$5000; t = 8 years; r = 6% = 0.06. We are trying to find the amount, A.

Annual compounding means n = 1, so we can use the second formula.

$$A = P(1 + r)^t$$

$$A = 5000(1 + 0.06)^8$$

$$A = 5000(1.5938)$$

$$A = \$7969.240$$

which rounded to the nearest dollar is \$7969.

To do the computation on a calculator, follow these steps:

ENTER	DISPLAY
1	1
+	1
.06	0.06
=	1.06
x^y or y^x	1.06
8	8
x	1.5938481
5000	5000
=	7969.2404

The display is 7969.2404, which we round to 7969. We can carry more significant figures if we need more precision.

Solution (b): The known quantities are the same except now we have monthly compounding, so n = 12 (there are 12 months in one year). Substituting into the compound interest formula we get:

$$A = P\left(1 + \frac{r}{n}\right)^{nt}$$

$$A = 5000\left(1 + \frac{0.06}{12}\right)^{12 \cdot 8}$$

$$A = 5000(1.005)^{96}$$

$$A = 5000(1.6141427)$$

$$A = 8070.7136$$

which rounded to the nearest dollar is \$8071.

To do the computation on a calculator, follow these steps:

ENTER	DISPLAY
.06	0.06
÷	0.06
12	12
=	0.005
+	0.005
1	1
=	1.005
x^y or y^x	1.005
(	1.005
12	12
x	12
8	8
)	96
x	1.6141427
5000	5000
=	8070.7136

The display is 8070.7136, which we round to 8071 or another number depending on our needed precision.

Solution (c): Here also the known quantities are the same. However, since interest is compounded continuously, we use the exponential formula:

$$A = P \cdot e^{rl}$$

$$A = 5000 \cdot e^{0.06 \cdot 8}$$

$$A = 5000 \cdot e^{0.48}$$

$$A = 5000 \cdot 1.6160744$$

$$A = 8080.372$$

which rounded to the nearest dollar is $8080.

To do the computation on a calculator, follow these steps:

ENTER	DISPLAY
.06	0.06
x	0.06
8	8
=	0.48
e^x	1.6160744
x	1.6160744
5000	5000
=	8080.372

The display is 8080.372, which we round to 8080 or other number depending on our needed precision. Some older calculators may not have an e^x button but they may have a ln or ln$x$ button. If so, we can hit INV and then ln$x$ in place of e^x. The exponential function, e^x, is the inverse of the natural logarithm function, ln$x$.

Comparing the three answers we see that they are close enough for estimating purposes.

annually	\$7969
monthly	\$8071
continuously	\$8080

Rounding to the nearest \$100 (or 2 significant figures) these values are 8000, 8100, and 8100. The exponential formula is the simplest to use and remember.

Example: Assuming an inflation rate of 3.5%, what is the expected cost of maintenance painting in 7 years if the cost today is \$24,000.
Solution: We first identify the known quantities. The cost today is analogous to the principal, so let P = \$24,000. The time, t = 7. The rate, r = 3.5% = 0.035. We are looking for the amount, A. Substituting into the formula:

$$A = P \cdot e^{rt}$$
$$A = 24{,}000 \cdot e^{0.035 \cdot 7}$$
$$A = 24{,}000 \cdot e^{0.245}$$
$$A = 24{,}000 \cdot 1.2776$$
$$A = 30662.9$$

Rounding to the nearest \$100 we find that the maintenance painting will cost about \$30,700 in 7 years at 3.5% inflation.

These problems can be worked in reverse to tell us how much we should invest now if we know how much we will need at a future date. If we know the amount in the future, what should be the principal? When discussing this type of problem, the principal is referred to as the present value, PV, and the amount is the future value, FV.

Making this change of notation and then rearranging the formula for annual compounding to solve for PV we get:

$$FV = PV \cdot (1 + r)^t$$

$$PV = \frac{FV}{(1 + r)^t}$$

OR

$$PV = FV(1 + r)^{-t}$$

A negative exponent indicates the reciprocal; for example,

$$a^{-n} = \frac{1}{a^n} \text{ (if } a \neq 0) \cdot$$

The exponential formula becomes:

$$FV = PV \cdot e^{rt}$$

$$PV = \frac{FV}{e^{rt}}$$

OR

$$PV = FV \cdot e^{-rt}$$

Example: Assume the rate of return on our investments is 5.5%. How much must we invest now to have $30,700 in seven years.

Solution: We know the rate, r = 5.5% = 0.055; the time, t = 7 years; the future value, FV= $30,700. We are looking for the amount to be invested now (the amount at time t = 0) or the present value, PV. We will solve the problem using either formula.

Assuming annual compounding we get:

$$PV = \frac{FV}{(1 + r)^t}$$

$$PV = \frac{30{,}700}{(1 + 0.055)^7}$$

$$PV = \frac{30{,}700}{1.454679}$$

$$PV = 21{,}104$$

To the nearest hundred, we would have to invest \$21,100 now at 5.5% to realize 30,700 in 7 years with annual compounding.

Assuming continuous compounding we get:

$$PV = \frac{FV}{e^{rt}}$$

$$PV = \frac{30{,}700}{e^{(0.055) \cdot (7)}}$$

$$PV = \frac{30{,}700}{1.4696}$$

$$PV = 20889.8$$

To the nearest hundred, we would have to invest \$20,900 now at 5.5% to realize \$30,700 in 7 years with continuous compounding.

If we keep only two significant figures and round to the nearest \$1000, both methods require an initial investment of \$21,000.

10.2 Comparing the Cost of Coating Systems

We often have to choose between an expensive painting system that will last for twenty years and an inexpensive painting system that will need maintenance after five or ten years. If we borrow money to install the best system, we will incur interest on the borrowed money. If we install the inexpensive system, we may be able to invest what is left so that when maintenance comes due we will have accrued enough interest to pay for the maintenance painting. Speculation is always tricky because no one can predict the future with certainty. Still, we can make intelligent guesses on expected rates and use mathematics to predict future amounts. This topic is discussed in *SSPC-TU 9 Estimating Costs in Protective Coatings Projects.*

Example: A zinc/epoxy/polyurethane system will cost $100,000 and last for 20 years without maintenance. A proprietary system will cost $60,000 initially and will require maintenance after 12 years. Today, the maintenance painting would cost $45,000. Assume inflation is 2.5%. Assume that the return on investment is 3.4%. Compare the cost of these two systems by computing the cost per year over a 20-year lifetime. Ignore all hidden costs due to factors such as down-time.
Solution: The cost per year for the zinc/epoxy/polyurethane system is simply the total cost ($100,000) divided by the lifetime (20 years) or $5000 per year.

The cost per year for the proprietary system requires some computation. The initial cost is given. In addition, we must find the cost of the maintenance painting in today's dollars. How much must be invested today to cover the maintenance costs in 12 years? First, we compute the actual cost of maintenance painting 12 years from now. Using the inflation rate of 2.5% ($r = 0.025$), a present value PV = $45,000, and time $t = 12$ years, we can find the future value by substituting into one of the formulas.

$$\begin{aligned} FV &= PV \cdot e^{rt} \\ FV &= 45{,}000 \cdot e^{(0.025)\cdot(12)} \\ FV &= 60{,}744 \end{aligned} \qquad \text{OR} \qquad \begin{aligned} FV &= PV(1+r)^{t} \\ FV &= 45{,}000(1+0.025)^{12} \\ FV &= 60{,}520 \end{aligned}$$

So maintenance painting in 12 years is expected to cost about $61,000.

The next question is how much do we invest now at 3.4% to have $61,000 in 12 years. We know the future value FV = 61,000, time $t = 12$, and rate $r = 3.4\% = 0.034$ and we are looking for the present value, PV.

Substituting into one of the formulas for present value:

$$PV = \frac{FV}{e^{rt}} \qquad \text{OR} \qquad PV = \frac{FV}{(1+r)^t}$$

$$PV = \frac{61{,}000}{e^{(0.034)(12)}} \qquad PV = \frac{61{,}000}{(1+0.034)^{12}}$$

$$PV = 40{,}564 \qquad PV = 40{,}840$$

So we must invest about \$41,000 now to be able to pay for the maintenance painting in 12 years. The total cost of the proprietary system in today's dollars is therefore the \$60,000 initial cost plus \$41,000 now to cover the future maintenance or a total of \$101,000. This is only slightly larger than the \$100,000 for the other system.

If we divide by the 20-year life time, we can find the cost per year as 101,000 ÷ 20 = 5050. This can be rounded to \$5000 per year. So the yearly cost of both systems is roughly \$5000. Based on cost alone, both systems are about the same. Other factors such as availability, service, company reputation, etc., will help form the final decision.

Appendix A: Rules and Formulas

Basic Algebra and Arithmetic

Rule for Adding Signed Numbers

- ☞ If the signs are the same, add the absolute values and keep the sign.
- ☞ If the signs are different, subtract the absolute values and keep the sign of the number with the larger absolute value.

Order of Operations Rule

1. First perform operations within grouping symbols such as parentheses.
2. Evaluate powers and roots.
3. Do all multiplication and division in order from left to right. (A division may come before a multiplication.)
4. Do addition and subtraction in order from left to right. (A subtraction may come before an addition.)

Rule for Subtracting Signed Numbers

- ☞ To subtract, add the opposite.

Rule for Multiplying and Dividing Signed Numbers

- ☞ When multiplying or dividing two numbers with the same sign, the answer is positive.
- ☞ When multiplying or dividing two numbers with opposite signs, the answer is negative.

A Fraction as a Multiplication

$$\frac{a}{b} = a \cdot \frac{1}{b} \quad \text{OR} \quad \frac{a}{b} = \frac{1}{b} \cdot a$$

A Fraction as a Division

$$\frac{a}{b} = a \div b \quad \text{OR} \quad \frac{a}{b} = b\overline{)a}$$

Basic Algebra and Arithmetic

Some Rules for Exponents

☞ When the exponent is not written, it is assumed to be 1.

$$6^1 = 6$$
$$b^1 = b$$

☞ If the exponent is zero, the quantity equals 1.

$$6^0 = 1$$
$b^0 = 1$ for any b except 0. (0^0 is not defined.)

☞ When multiplying numbers with the same base, add the exponents.

$$2^3 \cdot 2^4 = 2^7$$
$$b^m \cdot b^n = b^{m+n}$$

☞ When dividing numbers with the same base, subtract the exponents.

$$2^5 \div 2^3 = 2^2$$
$$b^m \div b^n = b^{m-n}$$

☞ When taking a power to a power, multiply the exponents.

$$(2^3)^2 = 2^{3 \cdot 2} = 2^6 = 64$$
$$(b^m)^n = b^{mn}$$

☞ If the exponent is negative, take the reciprocal then raise to the appropriate power.

$$b^{-n} = \frac{1}{b^n}$$
$$\left(\frac{a}{b}\right)^{-n} = \left(\frac{b}{a}\right)^n$$

☞ Zero to any positive power is zero. Zero to a negative power and zero to the zero power are not defined.

$$0^n = 0 \quad (n > 0)$$

Basic Algebra and Arithmetic

Steps for Solving Linear Equations

1. Remove parentheses using the distributive properties.
2. Combine like terms on each side.
3. Add and/or subtract to get all terms with variables on one side of the equation and all numbers on the other side.
4. Divide by the coefficient of the variable term.

Rules for Significant Figures

☞ All non-zero digits are significant

☞ Zeros between significant figures are significant

☞ All zeros at the beginning of a number are not significant

☞ Zeros at the end of a whole number are not significant

☞ Zeros at the end of a number after the decimal point are significant

Rules for Rounding

☞ If the next digit is less than 5, change that digit and those following to zero.

☞ If the next digit is 5 (followed by at least one non-zero digit) or greater, increase the digit to which we are rounding by one and change the following digits to zero.

☞ If the next digit is 5 followed by all zeros (if there are any more digits), leave the previous digit as is or add one to the previous digit so that the last digit of the rounded number is even.

☞ If the next digit is after the decimal point, it and the following digits are dropped.

Rule for Changing to and from a Percent

Since 100% = 1,

☞ To change TO a percent, multiply by 100%

☞ To change FROM a percent, divide by 100%

Basic Algebra and Arithmetic

Standard Deviation

$$s = \sqrt{\frac{\sum_{i=1}^{n} (x_i - x_{avg})^2}{n - 1}}$$

where

s = standard deviation
x_i = the data points
x_{avg} = the mean
n = number of data points

Geometry Formulas

A = area
P = perimeter

Square

$P = 4 \cdot s$
$A = s^2$

Rectangle

$P = 2L + 2W$

Rectangle
$A = L \bullet W$

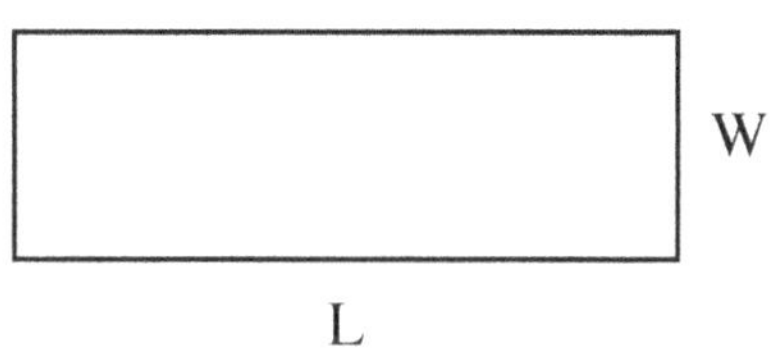

Geometry Formulas

Parallelogram

$A = b \cdot h$

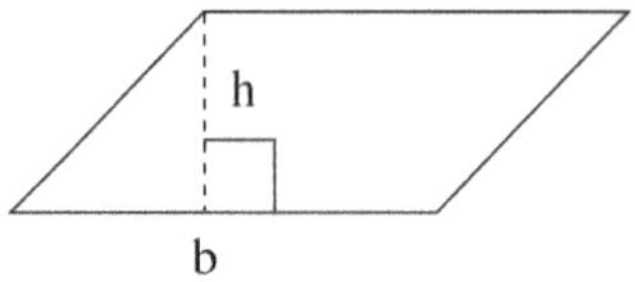

Trapezoid

area = (average base) · (height)

$$A = \frac{B + b}{2} \cdot h$$

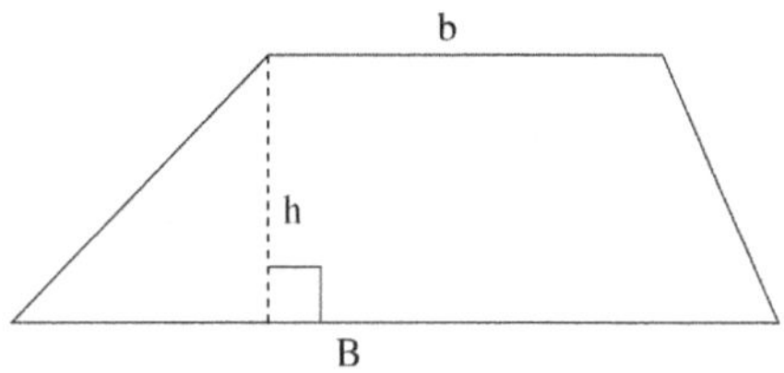

Triangle

$$A = \frac{1}{2} b \cdot h$$

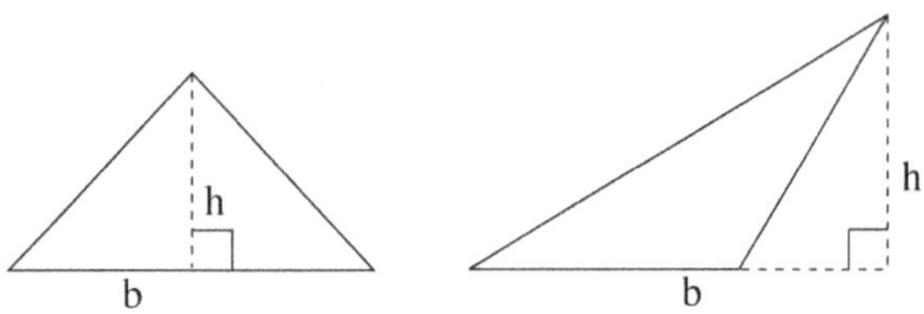

Circle

$d = 2r$

$C = 2\pi r$ $\qquad$ $C = \pi d$

$A = \pi r^2$ $\qquad$ $A = \pi \frac{d^2}{4}$

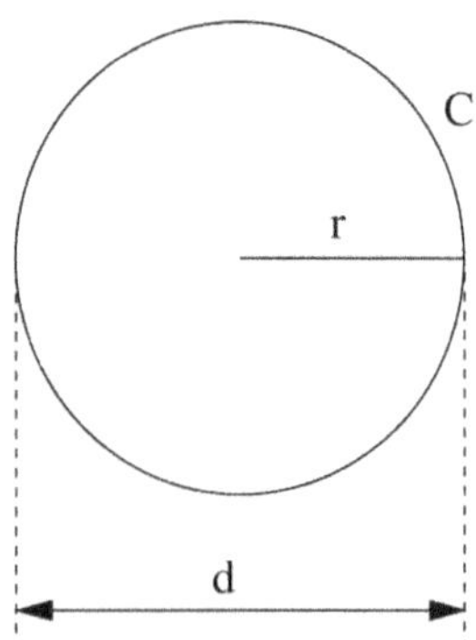

Geometry Formulas

General Area Formula

☞ The surface area (excluding the ends) of a regular object of length, L, and perimeter, P, is

$$A = P \cdot L$$

The cross section of the piece must be uniform so that the perimeter is the same at all points along the piece.

General Volume Formula

☞ The volume of a regular solid (e.g., a block or a cylinder) is computed by multiplying the area of the base times the height.

$$V = (\text{area of base}) \times (\text{height})$$

Rectangular Solid

$V = L \cdot W \cdot H$

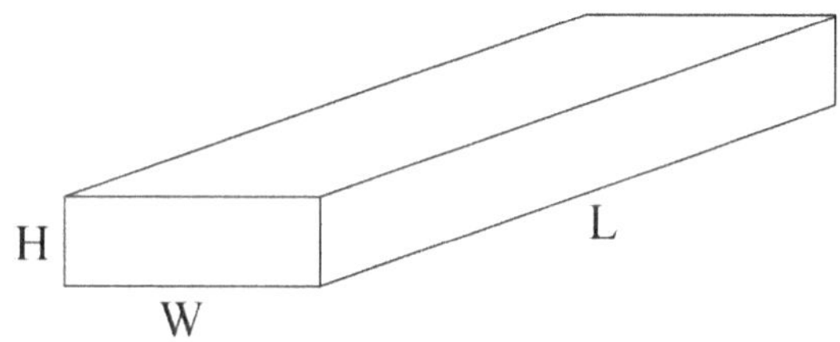

Cylinder

$V = \pi r^2 h$

$A_{side} = 2\pi rh$

$A_{top} = \pi r^2$

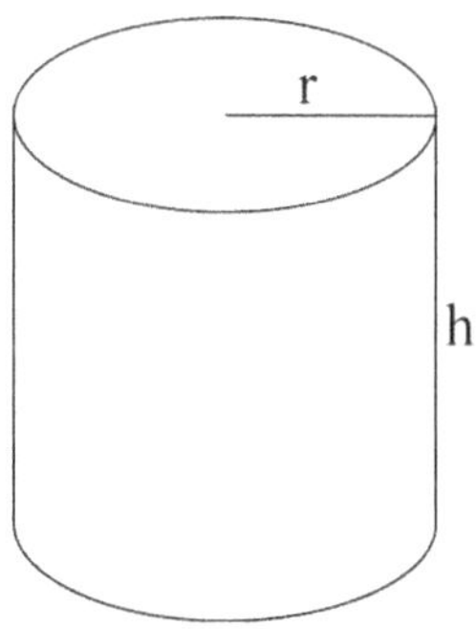

Chemistry Formulas and Definitions

Formulas for Conversion Between Fahrenheit and Celsius Temperatures

☞ If we know F and want to find C, then we use

$$C = \frac{F - 32}{1.8}$$

☞ If we know C and want to find F, then we use

$$F = 1.8C + 32$$

where:

F = the temperature reading in degrees Fahrenheit
C = the temperature reading in degrees Celsius

Definition of Density

☞ $D = \frac{W}{V}$ OR $D = \frac{M}{V}$

where

W = weight
V = volume
M = mass

Definition of Specific Gravity

$$\text{specific gravity} = \frac{\text{density of material}}{\text{density of water}}$$

$$\text{s.g.} = \frac{D}{D_{water}}$$

Chemistry Formulas and Definitions

Relationships Among Weight, Density, and Volume

$$D = \frac{W}{V} \qquad W = D \cdot V \qquad V = \frac{W}{D}$$

Formula for Concentration in ppm

$$\text{concentration (in ppm)} = \frac{\text{weight of dissolved salt (in } \mu\text{g)}}{\text{volume of solution (in mL)}}$$

assuming a dilute aqueous solution of density 1.0 g/mL.

Formula for Surface Concentration

$$\text{surface concentration } (\mu\text{g/cm}^2) = \frac{\text{extract concentration (in ppm)} \cdot \text{volume of extract (in mL)}}{\text{area of surface extracted (in cm}^2\text{)}}$$

Paint Computation Formulas

Formula Relating DFT, WFT, and Percent Volume Solids

$$\text{DFT} = \left(\begin{array}{c}\text{percent volume solids}\\ \text{expressed as a decimal or a fraction}\end{array}\right) \cdot \text{WFT}$$

or using single letters

☞ $$D = P_S \cdot W$$

The equation can be rearranged to solve for WFT

☞ $$W = \frac{D}{P_S}$$

The equation to solve for percent volume solids is

☞ $$P_S = \frac{D}{W}$$

where

D = dry film thickness
W = wet film thickness
P_S = percent volume solids (expressed as a decimal or a fraction)

Formula for Volume of Paint Needed

$$V = \frac{A \cdot D}{1604 \cdot P_S}$$

where

V = volume in gallons
A = area in ft^2
D = dry film thickness in mils
P_S = percent volume solids (expressed as a decimal or a fraction)
1604 = the conversion from mil · ft^2 to gal

Paint Computation Formulas

Relationship Among DFT, WFT, Percent Volume Solids, and Volume of Thinner

☞ To find DFT use

$$D = \left[\frac{V_S}{V_o + V_T}\right] \cdot W$$

☞ To find WFT use

$$W = \frac{D(V_o + V_T)}{V_S}$$

☞ To find V_S use

$$V_S = \frac{D(V_o + V_T)}{W}$$

where

V_o = original volume of paint (usually 1 gal)
V_T = volume of thinner added
V_S = volume of solids of unthinned paint from tech data sheet (can be measured in gal)
D = dry film thickness
W = wet film thickness

☞ If $V_o = 1$, V_S is numerically equal to the percent volume solids (in decimal form) of the unthinned paint.

☞ Corresponding formulas using **percent** thinner are given later in this section.

Paint Computation Formulas

Relationship Among DFT, WFT, Percent Volume Solids, and Percent of Thinner

☞ To find DFT use

$$D = \left[\frac{P_S}{1+P_T}\right] \cdot W$$

☞ To find WFT use

$$W = \frac{D(1+P_T)}{P_S}$$

☞ To find P_S use

$$P_S = \frac{D(1+P_T)}{W}$$

where

P_T = percent of thinner by volume added (expressed as a decimal or a fraction)

P_S = percent volume of solids of unthinned paint (expressed as a decimal or a fraction)

D = dry film thickness

W = wet film thickness

☞ Corresponding formulas using **volume** of thinner were given previously in this section.

Paint Computation Formulas

Formula for Computing Percent Volume Solids After Thinning

$$P_S = \frac{P_o}{1 + P_T}$$

where

P_S = % volume solids after thinning (expressed as a decimal or a fraction)

P_O = original percent volume solids (unthinned) (expressed as a decimal or a fraction)

P_T = percent of thinner by volume added

Formula for Spreading Rate (A/V) of Thinned Paint

$$\frac{A}{V} = \frac{1604 \cdot P_o}{D \cdot (1 + P_T)}$$

where

A = area in ft^2

V = volume in gal

D = dry film thickness in mils

P_o = percent volume solids of unthinned paint (expressed as a decimal or a fraction)

P_T = percent of thinner by volume added (expressed as a decimal or a fraction)

1604 = the conversion from mil · ft^2 to gal

Definition of Transfer Efficiency

$$\text{transfer efficiency} = \frac{\text{volume of coating applied to surface}}{\text{volume of coating used}}$$

OR

$$\text{transfer efficiency} = \frac{\text{mass (weight) of coating applied to surface}}{\text{mass (weight) of coating used}}$$

Paint Computation Formulas

Amount of Paint to be Ordered Beyond the Theoretical Value When Painting a Flat Surface

APPLICATION METHOD	ADDITIONAL PAINT NEEDED
brush	5%
roll	10%
spray	20 – 30%

Definition of Pigment Volume Concentration (PVC)

$$PVC = \frac{\text{volume of pigment}}{\text{volume of total solids}} = \frac{V_p}{V_p + V_b}$$

where

V_p = volume of pigment
V_b = volume of binder

Flow Rate Formula

flow rate = (cross sectional area) · (velocity of material)

Or using letters,

$$Q = A \cdot v$$

where

Q = flow rate
A = cross sectional area
v = velocity of material

Paint Computation Formulas

Formula to Compute the Flow Capacity of a Dehumidifier (U.S. Customary Version)

$$X = \frac{(V_i)(RAC)}{60}$$

where

X = air-flow capacity in ft^3/min

V_i = the internal volume of the space minus the volume of any obstructions in ft^3

RAC = the required air changes per hour

60 = the conversion factor from hours to minutes

Formula to Compute the Flow Capacity of a Dehumidifier (Metric Version)

$$X = (V_i)(RAC)$$

where

X = air-flow capacity in m^3/h

V_i = the internal volume of the space minus the volume of any obstructions in m^3

RAC = the required air changes per hour

Money Formulas

Compound Interest Formula

$$A = P\left(1 + \frac{r}{n}\right)^{nt}$$

where

A = amount after t years
P = principal (amount at time zero)
r = annual rate expressed as a decimal
n = number of compounding periods per year
(n = 4 for quarterly compounding)
t = time in years
r/n = rate per compounding period
nt = total number of periods in t years

Formula for Annual Compounding

$A = P(1 + r)^t$ OR $FV = PV(1 + r)^t$

where

A = amount after t years
P = principal (amount at time zero)
r = annual rate expressed as a decimal
t = time in years
FV = future value
PV = present value

Money Formulas

Formula for Continuous Compounding

$$A = P e^{rt}$$

OR

$$FV = PV \cdot e^{rt}$$

$$PV = \frac{FV}{e^{rt}}$$

OR

$$PV = FV \cdot e^{-rt}$$

where

A = amount after t years
P = principal (amount at time zero)
r = annual rate expressed as a decimal
t = time in years
e = base of the natural logarithms (≈ 2.71828)
FV = future value
PV = present value

Appendix B: Conversion Factors

Conversion Factors

Refer to IEEE/ASTM SI 10 for exact conversion factors. Non-exact factors are given to at least four significant figures. Only factors that may relate to the coating industry are listed. Gallons are U.S. gallons.

SI Metric Prefixes

PREFIX	SYMBOL	MULTIPLIER	POWER OF 10 FORM
deka	da	10	10^1
hecto	h	100	10^2
kilo	k	1000	10^3
mega	M	1,000,000	10^6
giga	G	1,000,000,000	10^9
deci	d	0.1	10^{-1}
centi	c	0.01	10^{-2}
milli	m	0.001	10^{-3}
micro	µ	0.000 001	10^{-6}
nano	n	0.000 000 001	10^{-9}

Length

SI ⇔ SI	U.S. ⇔ U.S.	U.S. ⇔ SI	SI ⇔ U.S.
1 µm = 0.001 mm	1 mil = 0.001 inch	1 mil = 25.4 µm	1 µm = 0.039 37 mil
1µm = 0.000 001 m	1 inch = 1000 mils	1 inch = 25.4 mm	1 µm = 0.000 039 37 inch
1 mm = 0.001 m	1 ft = 12 inch	1 inch = 2.54 cm	1 mm = 39.37 mil
1 mm = 0.1 cm	1 yd = 36 inch	1 ft = 30.48 cm	1 mm = 0.039 37 inch
1 cm = 10 mm	1 yd = 3 ft	1 ft = 0.3048 m	1 cm = 0.393 7 inch
1 cm = 0.01 m	1 mile = 5280 ft	1 yd = 0.9144 m	1 m = 39.37 inch
1 m = 1000 mm	1 mile = 1760 yd	1 mile = 1609 m	1 km = 3281 ft
1 m = 100 cm		1 mile = 1.609 km	1 km = 0.621 4 mile
1 km = 1000 m			

Appendix B: Conversion Factors

Weight (Mass)

SI ⇔ SI	U.S. ⇔ U.S.	U.S. ⇔ SI	SI ⇔ U.S.
1 μg = 0.001 mg	1 gr = 0.002 857 oz	1 gr = 64.80 mg	1 μg = 0.000 015 43 gr
1 mg = 1000 μg	1 oz = 437.5 gr	1 oz = 28.35 g	1 mg = 0.015 43 gr
1 g = 1,000,000 μg	1 oz = 0.062 5 lb	1 lb = 453.6 g	1 g = 0.035 27 oz
1 g = 1000 mg	1 lb = 16 oz	1 lb = 0.453 6 kg	1 g = 0.002 205 lb
1 kg = 1000 g	1 ton = 2000 lb	1 ton = 907.2 kg	1 kg = 2.205 lb
1 tonne = 1000 kg		1 ton = 0.907 2 tonne	1 tonne = 1.102 ton

Area

SI ⇔ SI	U.S. ⇔ U.S.	U.S. ⇔ SI	SI ⇔ U.S.
1 cm^2 = 100 mm^2	1 ft^2 = 144 in^2	1 in^2 = 6.452 cm^2	1 cm^2 = 0.155 0 in^2
1 m^2 = 10,000 cm^2	1 yd^2 = 9 ft^2	1 ft^2 = 929.0 cm^2	1 m^2 = 1550 in^2
		1 ft^2 = 0.092 90 m^2	1 m^2 = 10.76 ft^2
		1 yd^2 = 0.8361 m^2	

Volume

SI ⇔ SI	U.S. ⇔ U.S.	U.S. ⇔ SI	SI ⇔ U.S.
1 mL = 1 cm^3 = 1 cc	1 in^3 = 1.805 oz	1 in^3 = 16.39 cm^3	1 cm^3 = 0.061 02 in^3
1 mL = 0.001 L	1 oz = 0.031 25 qt	1 in^3 = 16.39 mL	1 mL = 0.061 02 in^3
1 L = 1000 mL	1 pt = 16 oz	1 oz = 29.57 mL	1 L = 61.02 in^3
1 L = 1000 cm^3	1 qt = 32 oz	1 qt = 1.101 L	1 L = 0.035 31 ft^3
1 L = 0.001 m^3	1 qt = 2 pt	1 gal = 3.785 L	1 L = 2.113 pt
1 m^3 = 1000 L	1 gal = 231.0 in^3	1 ft^3 = 28,320 cm^3	1 L = 1.057 qt
	1 gal = 0.133 7 ft^3	1 ft^3 = 0.028 32 m^3	1 L = 0.264 2 gal
	1 gal = 4 qt	1 ft^3 = 28.32 L	1 m^3 = 61,023 in^3
	1 ft^3 = 1728 in^3	1 yd^3 = 0.764 6 m^3	1 m^3 = 35.31 ft^3
	1 ft^3 = 7.480 52 gal	1 yd^3 = 764.6 L	1 m^3 = 1.308 yd^3
	1 ft^3 = 0.1111 yd^3		1 m^3 = 264.2 gal
	1 yd^3 = 27 ft^3		

Density

SI ⇔ SI	U.S. ⇔ U.S.	U.S. ⇔ SI	SI ⇔ U.S.
1 g/cm^3 = 1000 kg/m^3 1 g/L = 1 kg/m^3	1 lb/ft^3 = 0.133 7 lb/gal 1 lb/gal = 7.480 5 lb/ft^3	1 lb/ft^3 = 16.018 kg/m^3 1 lb/ft^3 = 16.018 g/L 1 lb/gal = 119.8 kg/m^3 1 lb/gal = 119.8 g/L 1 oz/gal = 7.489 g/L	1 kg/m^3 = 0.062 43 lb/ft^3 1 g/L = 0.008 345 lb/gal

Pressure

SI ⇔ SI	U.S. ⇔ U.S.	U.S. ⇔ SI	SI ⇔ U.S.
1 Pa = 1 N/m^2 1 millibar = 100 Pa 1 mm Hg = 133.3 Pa 1 torr = 133.3 Pa	1 atm = 14.70 lb/in^2 1 atm = 14.70 psi 1 psi = 144 lb/ft^2 1 lb/ft^2 = 0.006 944 lb/in^2	1 in Hg = 3.386 kPa 1 in water = 249.1 Pa 1 atm = 101.3 kPa 1 psi = 6.895 kPa 1 lb/ft^2 = 992.9 kPa 1 lb/ft^2 = 0.992 9 MPa	1 kPa = 0.1450 psi 1 kPa = 20.88 lb/ft^2 1 MPa = 145.0 psi

Miscellaneous Conversion Factors

☞ **FLOW RATE**

1L/min = 0.03531 CFM
1 L/s = 2.119 CFM
1 CFM = 0.000 0471 9 m^3/s
1 CFM = 0.471 9 L/s
1 CFM = 1 ft^3/min

☞ **CONSUMPTION** (per unit area)

1 lb/ft^2 = 4.882 kg/m^2
1 kg/m^2 = 0.204 8 lb/ft^2

☞ **CONSUMPTION** (per unit time)

1 lb/min = 0.007 560 kg/s
1 lb/h = 0.453 6 kg/h
1 kg/s = 132.2 lb/min
1 kg/h = 2.205 lb/h

☞ **CONCENTRATION**

1 ppm = 1 µg/mL = 1 mg/L
1 ppm = 0.058 42 gr/gal
1 gr/gal = 17.118 mg/L
1 lb/gal = 119.8 g/L
1 g/L = 0.008345 lb/gal

☞ **WATER VAPOR TRANSMISSION** (Permanence)

1 perm (0°C) = 5.721 x 10^{-11} $kg/(Pa \cdot s \cdot m^2)$
1 perm = 1 $gr/ft^2/h$/inch Hg

☞ **TORQUE**

1 inch·lb = 0.112 98 N·m

Abbreviations

U.S. CUSTOMARY	
SYMBOL	**NAME**
s	second
min	minute
h	hour
mil	mil (0.001 inch)
in	inch
ft	foot
yd	yard
mi	mile
gr	grain
oz	ounce
lb	pound
pt	pint
qt	quart
gal	gallon
in Hg	inch of mercury
in water	inch of water
psi	pound per square inch
atm	atmosphere
CFM	cubic feet per minute
°F	degree Fahrenheit

SI (METRIC)	
SYMBOL	**NAME**
s	second
min	minute
h	hour
µm	micrometer (micron)
m	meter
g	gram
kg	kilogram
L	liter
Pa	pascal
N	newton
mm Hg	millimeter of mercury
cm Hg	centimeter of mercury
Pa	pascal
J	joule
S	siemens
°C	degree celsius
K	kelvin
W	watt

Appendix C: Temperature Conversion Charts

Temperature Conversion Charts

Conversion from Degrees Celsius to Degrees Farenheit

°C	°F	°C	°F	°C	°F	°C	°F	°C	°F
-10	14	20	68	50	122	80	176	100	212
-9	16	21	70	51	124	81	178	125	257
-8	18	22	72	52	126	82	180	150	302
-7	19	23	73	53	127	83	181	175	347
-6	21	24	75	54	129	84	183	200	392
-5	23	25	77	55	131	85	185	225	437
-4	25	26	79	56	133	86	187	250	482
-3	27	27	81	57	135	87	189	275	527
-2	28	28	82	58	136	88	190	300	572
-1	30	29	84	59	138	89	192	325	617
0	32	30	86	60	140	90	194	350	662
1	34	31	88	61	142	91	196	375	707
2	36	32	90	62	144	92	198	400	752
3	37	33	91	63	145	93	199	425	797
4	39	34	93	64	147	94	201	450	842
5	41	35	95	65	149	95	203	475	887
6	43	36	97	66	151	96	205	500	932
7	45	37	99	67	153	97	207	525	977
8	46	38	100	68	154	98	208	550	1022
9	48	39	102	69	156	99	210	575	1067
10	50	40	104	70	158	100	212	600	1112
11	52	41	106	71	160	101	214	625	1157
12	54	42	108	72	162	102	216	650	1202
13	55	43	109	73	163	103	217	675	1247
14	57	44	111	74	165	104	219	700	1292
15	59	45	113	75	167	105	221	725	1337
16	61	46	115	76	169	106	223	750	1382
17	63	47	117	77	171	107	225	775	1427
18	64	48	118	78	172	108	226	800	1472
19	66	49	120	79	174	109	228	825	1517

Degrees Fahrenheit were rounded to the nearest degree.
Fahrenheit temperatures are computed using $F = 1.8C + 32$.

Conversion from Degrees Farenheit to Degrees Celsius

°F	°C	°F	°C	°F	°C	°F	°C	°F	°C
-10	-23	20	-7	50	10	80	27	100	38
-9	-23	21	-6	51	11	81	27	150	66
-8	-22	22	-6	52	11	82	28	200	93
-7	-22	23	-5	53	12	83	28	250	121
-6	-21	24	-4	54	12	84	29	300	149
-5	-21	25	-4	55	13	85	29	350	177
-4	-20	26	-3	56	13	86	30	400	204
-3	-19	27	-3	57	14	87	31	450	232
-2	-19	28	-2	58	14	88	31	500	260
-1	-18	29	-2	59	15	89	32	550	288
0	-18	30	-1	60	16	90	32	600	316
1	-17	31	-1	61	16	91	33	650	343
2	-17	32	0	62	17	92	33	700	371
3	-16	33	1	63	17	93	34	750	399
4	-16	34	1	64	18	94	34	800	427
5	-15	35	2	65	18	95	35	850	454
6	-14	36	2	66	19	96	36	900	482
7	-14	37	3	67	19	97	36	950	510
8	-13	38	3	68	20	98	37	1000	538
9	-13	39	4	69	21	99	37	1050	566
10	-12	40	4	70	21	100	38	1100	593
11	-12	41	5	71	22	101	38	1150	621
12	-11	42	6	72	22	102	39	1200	649
13	-11	43	6	73	23	103	39	1250	677
14	-10	44	7	74	23	104	40	1300	704
15	-9	45	7	75	24	105	41	1350	732
16	-9	46	8	76	24	106	41	1400	760
17	-8	47	8	77	25	107	42	1450	788
18	-8	48	9	78	26	108	42	1500	816
19	-7	49	9	79	26	109	43	1550	843

Degrees celsius were rounded to the nearest degree.

Celsius temperatures are computed using $C = \frac{F - 32}{1.8}$

Appendix D:
Theoretical Spreading Rates

Theoretical Spreading Rates

Theoretical Spreading Rates (ft^2/gal)
at Several Dry Film Thicknesses and Volume Solids

% Vol. Solids	Dry Film Thickness (DFT)							
	1 mil	**2 mils**	**3 mils**	**4 mils**	**5 mils**	**6 mils**	**8 mils**	**10 mils**
30	481	241	160	120	96	80	60	48
35	561	281	187	140	112	94	70	56
40	642	321	214	160	128	107	80	64
45	722	361	241	180	144	120	90	72
50	802	401	267	201	160	134	100	80
55	882	441	294	221	176	147	110	88
60	962	481	321	241	192	160	120	96
65	1043	521	348	261	209	174	130	104
70	1123	561	374	281	225	187	140	112
72	1155	577	385	289	231	192	144	115
74	1187	593	396	297	237	198	148	119
76	1219	610	406	305	244	203	152	122
78	1251	626	417	313	250	209	156	125
80	1283	642	428	321	257	214	160	128
82	1315	658	438	329	263	219	164	132
84	1347	674	449	337	269	225	168	135
86	1379	690	460	345	276	230	172	138
88	1412	706	471	353	282	235	176	141
90	1444	722	481	361	289	241	180	144
92	1476	738	492	369	295	246	184	148
94	1508	754	503	377	302	251	188	151
96	1540	770	513	385	308	257	192	154
98	1572	786	524	393	314	262	196	157
100	1604	802	535	401	321	267	201	160

Table values were computed from the formula

$$\text{Spreading Rate} = \frac{A}{V} = \frac{1604 \cdot P_S}{D}$$

where

A = area in ft^2
V = volume in gal
D = dry film thickness in mils
P_S = %volume solids (expressed as a decimal or a fraction)
1604 = the conversion from mil·ft^2 to gal

Appendix E:
Dew Point and Relative Humidity Charts

Dew Point Charts

Dew Point Temperature in Degrees Fahrenheit (Pressure = 30.0 in Hg)

Air Temp (°F) t	Temperature Depression of Wet Bulb Thermometer (t-t')									
	1.0	2.0	3.0	4.0	5.0	6.0	7.0	8.0	9.0	10.0
30	27	25	21	18	14	8	2			
31	28	26	23	19	15	10	4			
32	30	27	24	21	17	12	7			
33	31	28	25	22	18	14	9	2		
34	32	29	26	23	20	16	11	5		
35	33	30	28	25	21	17	13	7		
36	34	31	29	26	23	19	15	10	3	
37	35	32	30	27	24	21	17	12	6	
38	36	33	31	28	25	22	18	14	8	1
39	37	34	32	29	27	23	20	16	11	4
40	38	35	33	30	28	25	21	18	13	7
41	39	36	34	31	29	26	23	19	15	10
42	40	38	35	33	30	27	24	21	17	12
43	41	39	36	34	31	28	25	22	19	14
44	42	40	37	35	32	30	27	24	20	16
45	43	41	38	36	34	31	28	25	22	18
46	44	42	40	37	35	32	29	27	23	20
47	45	43	41	38	36	33	31	28	25	21
48	46	44	42	40	37	35	32	29	26	23
49	47	45	43	41	38	36	33	30	28	24
50	48	46	44	42	40	37	34	32	29	26
51	49	47	45	43	41	38	36	33	30	27
52	50	48	46	44	42	40	37	34	32	29
53	51	49	47	45	43	41	38	36	33	30
54	52	50	48	46	44	42	40	37	34	32
55	53	51	50	48	45	43	41	38	36	33
56	54	53	51	49	47	44	42	40	37	34
57	55	54	52	50	48	46	43	41	39	36
58	56	55	53	51	49	47	45	42	40	37
59	57	56	54	52	50	48	46	44	41	39
60	58	57	55	53	51	49	47	45	43	40
61	59	58	56	54	52	50	48	46	44	42
62	60	59	57	55	53	52	50	47	45	43
63	61	60	58	56	55	53	51	49	47	44
64	62	61	59	57	56	54	52	50	48	46
65	63	62	60	59	57	55	53	51	49	47
66	64	63	61	60	58	56	54	52	50	48
67	65	64	62	61	59	57	55	53	52	49
68	67	65	63	62	60	58	57	55	53	51
69	68	66	64	63	61	59	58	56	54	52

Appendix E: Dew Point and Relative Humidity

Dew Point Temperature in Degrees Fahrenheit (Pressure = 30.0 in Hg)

Air Temp (°F) t	Temperature Depression of Wet Bulb Thermometer (t-t')									
	1.0	2.0	3.0	4.0	5.0	6.0	7.0	8.0	9.0	10.0
70	69	67	65	64	62	61	59	57	55	53
71	70	68	67	65	63	62	60	58	56	54
72	71	69	68	66	64	63	61	59	58	56
73	72	70	69	67	66	64	62	60	59	57
74	73	71	70	68	67	65	63	62	60	58
75	74	72	71	69	68	66	64	63	61	59
76	75	73	72	70	69	67	66	64	62	60
77	76	74	73	71	70	68	67	65	63	62
78	77	75	74	72	71	69	68	66	64	63
79	78	76	75	73	72	70	69	67	66	64
80	79	77	76	74	73	72	70	68	67	65
81	80	78	77	75	74	73	71	70	68	66
82	81	79	78	77	75	74	72	71	69	67
83	82	80	79	78	76	75	73	72	70	69
84	83	81	80	79	77	76	74	73	71	70
85	84	82	81	80	78	77	75	74	72	71
86	85	83	82	81	79	78	76	75	73	72
87	86	84	83	82	80	79	78	76	75	73
88	87	85	84	83	81	80	79	77	76	74
89	88	86	85	84	82	81	80	78	77	75
90	89	87	86	85	83	82	81	79	78	76
91	90	88	87	86	85	83	82	80	79	78
92	91	89	88	87	86	84	83	81	80	79
93	92	90	89	88	87	85	84	83	81	80
94	93	92	90	89	88	86	85	84	82	81
95	94	93	91	90	89	87	86	85	83	82
96	95	94	92	91	90	88	87	86	84	83
97	96	95	93	92	91	89	88	87	85	84
98	97	96	94	93	92	90	89	88	87	85
99	98	97	95	94	93	92	90	89	88	86
100	99	98	96	95	94	93	91	90	89	87
101	100	99	97	96	95	94	92	91	90	88
102	101	100	98	97	96	95	93	92	91	89
103	102	101	99	98	97	96	94	93	92	91
104	103	102	100	99	98	97	95	94	93	92
105	104	103	101	100	99	98	96	95	94	93
106	105	104	102	101	100	99	98	96	95	94
107	106	105	103	102	101	100	99	97	96	95
108	107	106	104	103	102	101	100	98	97	96
109	108	107	105	104	103	102	101	99	98	97
110	109	108	106	105	104	103	102	100	99	98

Data taken from: Vapor Pressure, Relative Humidity, and Temperature of the Dew Point, C. F. Marvin, U. S. Department of Commerce, Weather Bureau.

Relative Humidity Chart

Relative Humidity, Percent (Pressure = 30.0 in Hg)

Air Temp (°F) t	Temperature Depression of Wet Bulb Thermometer (t-t')														
	1.0	2.0	3.0	4.0	5.0	6.0	7.0	8.0	9.0	10.0	11.0	12.0	13.0	14.0	15.0
30	89	78	67	56	46	36	26	16	6						
32	89	79	69	59	49	39	30	20	11	2					
34	90	81	71	62	52	43	34	25	16	8					
36	91	82	73	64	55	46	38	29	21	13	5				
38	92	83	75	66	58	50	42	33	25	17	10	2			
40	92	83	75	68	60	52	45	37	29	22	15	7	0		
42	92	85	77	69	62	55	47	40	33	26	19	13	5		
44	93	85	78	71	63	56	49	43	36	30	23	16	10	4	
46	93	86	79	72	66	58	52	45	39	32	26	20	14	8	2
48	93	86	79	73	67	60	54	47	41	35	29	23	18	12	7
50	93	87	81	74	67	61	55	49	43	38	32	27	21	16	10
52	94	87	81	75	69	63	57	51	46	40	35	29	24	19	14
54	94	88	82	76	70	64	59	53	48	42	37	32	27	22	17
56	94	88	82	76	71	65	60	55	50	44	39	34	30	25	20
58	94	88	83	77	72	66	61	56	51	46	41	37	32	27	23
60	94	89	83	78	73	68	63	58	53	48	43	39	34	30	26
62	94	89	84	79	74	69	64	59	54	50	45	41	36	32	28
64	95	90	84	79	74	70	65	60	56	51	47	43	38	34	30
66	95	90	85	80	75	71	66	61	57	53	48	44	40	36	32
68	95	90	85	80	76	71	67	62	58	54	50	46	42	38	34
70	95	90	86	81	77	72	68	64	59	55	51	48	44	40	36
72	95	91	86	82	77	73	69	65	61	57	53	49	45	42	38
74	95	91	86	82	78	74	69	65	61	58	54	50	47	43	39
76	96	91	87	82	78	74	70	66	62	59	55	51	48	44	41
78	96	91	87	83	79	75	71	67	63	60	56	53	49	46	43
80	96	91	87	83	79	75	72	68	64	61	57	54	50	47	44
82	96	92	88	84	80	76	72	69	65	61	58	55	51	48	45
84	96	92	88	84	80	76	73	69	66	62	59	56	52	49	46
86	96	92	88	84	81	77	73	70	66	63	60	57	53	50	47
88	96	92	88	85	81	77	74	70	67	64	61	57	54	51	48
90	96	92	89	85	81	78	74	71	68	65	61	58	55	52	49
92	96	92	89	85	82	78	75	72	68	65	62	59	56	53	50
94	96	93	89	85	82	79	75	72	69	66	63	60	57	54	51
96	96	93	89	86	82	79	76	73	69	66	63	61	58	55	52
98	96	93	89	86	83	79	76	73	70	67	64	61	58	56	53
100	96	93	89	86	83	80	77	73	70	68	65	62	59	56	54
102	96	93	90	86	83	80	77	74	71	68	65	62	60	57	55
104	97	93	90	87	83	80	77	74	71	69	66	63	60	58	55
106	97	93	90	87	84	81	78	75	72	69	66	64	61	58	56
108	97	93	90	87	84	81	78	75	72	70	67	64	62	59	57
110	97	93	90	87	84	81	78	75	73	70	67	65	62	60	57

Data taken from: Vapor Pressure, Relative Humidity, and Temperature of the Dew Point, C. F. Marvin, U. S. Department of Commerce, Weather Bureau.

INDEX

Abrasive use estimates 126–127
Absolute humidity 156
Absolute temperature scale 45
Absolute value 2
Absolute zero 155
Acceptance criteria 154–155
Accuracy versus precision 149
Addition 2–3
Air flow 142–143
Area 54, 126
Arithmetic
  addition 2–3
  division 4–5
  formulas 190–193
  fractions 7–15
  linear equations 21–24
  multiplication 4–5
  order of operations 5–6
  roots 18
  signed numbers 1–5
  subtraction 3
Area measurements 47–48
ASTM
  D 1475 150, 154
  E 691 150
  SI 10 34, 37, 208–210

Bias statements explained 154–155
Binder 97, 132

Caliper measurements 149–150
Capacity
  dehumidification flow 146
  liters 59
Chemistry
  rules and formulas 196–197
Chloride concentrations 137–142

Circular dimensions 51–52
Circumference 51–52
Coating system cost comparisons 168–170
Coatings
 job estimates 124–129
 moisture-cure polyurethanes 155
 mixing ratios 28–29
 zinc-rich 155
Composition (paint) 97, 98, 132–133
Computing
 area 7, 16
 cleaning time 84–85
 coating system costs 168–170
 compound interest 161
 costs 161–170
  and discounts 82–84
  dehumidifier flow capacity 146, 203
 dew point 155–159
 DFT
  from spreading rate 119–121
  measurements 88, 94
 interest 161–166, 204–205
 maintenance painting costs 166–167
 paint coverage 102–104
 paint waste 64–65
 pigment volume concentration 136
 proportion 25–27
 quantities based on paint composition 132–133
 relative humidity 155–159
 salt concentrations 139–142
 spreading rate 135–136
 VOC 26, 135
 volume solids 97–98, 114–116
 volume flow rate 142–143
Condensation 157
Confidence
 intervals 94
 levels 155
Containment 144

Converting
- gallons to liters 57–59
- meters to micrometers 73
- percents 79

Conversion factors 208–212
Cylinder volume 57–58

Data range 90–92
Decimals 15–17
- notation 73

Dehumidification 145–148, 155
- equipment flow capacity 146, 203

Deionized water 139–140
Density 134
- defined 60–61

Dew point
- charts 220–221
- computing 155–159

Diameter 51–52
Dimensional measurements
- defined 33

Dispersion 90
Division 4–5, 8
Dry bulb thermometers 157
Dry film thickness
- and spreading rates 218
- and SSPC PA 2 150–151
- computing measurements 88, 94, 100–101, 135
- gauges 82
- relationship to wet 100
- spot measurements 87, 151–152

Dust collector 144

Errors
- random versus systematic 150

Estimating
- costs 161–170
- maintenance painting 166–167
- materials 124–129
- shim thickness 69

Estimating, cont.
VOC content 69
work hours 68
Equations
defined 21
Evaporation 157
Exponents
in computing area and volume 16
Extracting methods (salts) 139–140

Film thickness 82, 87–88, 94, 100
and percent of thinner 117
relationship to volume solids and thinner 111
Flow rate formulas 142–143
Formulas
area 54
chemistry 196–197
dehumidifier flow capacity 146, 203
DFT, WFT, and volume solids 102
film thickness and percent of thinner 117
flow rate formulas 142–143
geometry 193–195
interest 161–166, 204–205
paint 40–41, 198–203
pigment volume concentration 131–132
salt concentrations 139–142
spreading rate of thinned paint 114–116
volume 55–56
Fractions
adding or subtracting 11–12
converting 8–9
dividing 8–9, 11
improper 8
multiplying 7–8, 10
reducing 10
using a calculator 13–14

Geometric measurements 50–51
Geometry rules and formulas 193–195

Grain
  defined 156
Gravity
  specific 62–63
Gauge
  readings of spot measurements 151
  Type-1 pull off 150

H-beams 53–54
Humidity 156, 222

IEEE/ASTM SI 10 34, 37, 208–210
Interest 161–166, 204–205
Iron oxide 133

Kelvin temperature scale 155

Laminar flow 142–143

Mass
  defined 60
Maintenance painting costs 166–167
Materials estimating 124–129
Mean 87–88
Median 87–88
Measurements systems
  defined 33
Measuring
  area 47–49
  average value 150
  circumference 51–52
  cylinders 57–58
  DFT 88, 94
  diameter 51–52
  geometric shapes 50–51
  H-beams 53–54
  metric and U.S. conversion factors 208–212
  perimeter 50–51
  pipe circumference 55
  radius 51–52

Measuring cont.
- spot readings 151
- tanks 52–53, 57–58

Metric system
- and U.S. conversion factors 208–212
- defined 33–35

Mixing ratios (for coatings) 28–29
Mode 88
Multiplication 4–5, 7–8
- distributive property 23

National Institute of Standards and Technology 33
Negative
- exponents 17–18
- numbers 17

Numbers
- rounding 67–69

Paint
- composition 97–98, 132–133
- coverage
  - formula 40–41
  - volume 103–105
- density 61–62
- effect of thinning 106–108
- estimating 124–129
- formulas 198–203
- waste 64–65

Percents
- converting fractions to decimals 76–77
- rules of use 79

Perimeter 50–51
Pi
- defined 51

Pigment 25, 97–98
- volume concentration 131–132, 136

Pipe circumference 55
Precision
- statements explained 154–155
- versus accuracy 149

Primer 127
Proportion 25–17
Psychrometer 157
Psychrometric tables 158

Quality assurance
  testing for density 154–155

Radius 51–52
Ratio 25–26, 28
  of pigment to vehicle 25
  of weight to volume 60–61
Reciprocals 11
Relative humidity 156, 222
Resin 97
Roots 18
Rust retardation 155

Salt concentration 136–142
Scientific notation 72–75
Shim thickness 69
Significant figures
  and job estimates 67–69
  rules for use 71
Signed numbers 1–5
Sling psychrometer 157
Solids 97–98
Solvent 97
Specific gravity 62–63
Spreading rate 132
  and volume solids/DFT 218
  computing DFT from 119–121
  effect of thinning 106–108
  formula 114–116
SSPC
  Guide 15 136
  PA 1 44
  PA 2 94, 150–153
  Paint 25 132–133, 135
  SP 6 125

SSPC cont.
- TR 3/NACE 6A192 146
- TU 4 136
- TU 9 169
- VIS 2 60

Standard deviations 155
Statistical mean 87–88
Steel
- density and weight 63–64
- surface temperature 155

Surface
- concentration of salts 136
- profile 157

Subtraction 3–4

Tanks 52–53
- volume measurements 57–58
- with dehumidification equipment 147

Temperature
- and dew point computations 155–159
- defined 155
- conversions 42–45, 214–215

Thermometers 157
Thinning
- effect of spreading rate 106–108
- effect on VOC content 116–117

Thinner 135
Top coats 128–129
Transfer efficiency 121–124
Type-1 pull-off gauge 150

Unit conversions 35–39
U.S. and metric conversion factors 208–212

Vehicle 25, 97–98
VOC
- computations 26, 68–69
- content 38–39, 132
  - and spreading rate 135–136
  - effect on thinning 116–117

Volume
concentration of salts 137–138
formula 55–56
paint coverage 103–105
solids 97–98, 107
and spreading rates 218
percent after thinning 114–116

Waste 64–65
Waterjetting 36–37, 64
Weight
defined 60
units (grain) 156
Wet bulb thermometers 157
Wet film thickness
relationship to dry 100

Zinc oxide 133–134
Zinc-rich primer 152

GLOSSARY

Abrasive. (1) A material used for wearing away a surface by rubbing. (2) A fine, granulated material used for blast cleaning. Abrasive particles of controlled mesh sizes are propelled by compressed air, water, or centrifugal force to clean and roughen a surface. Blast cleaning abrasives often are simply referred to as metallic or non-metallic and as shot- or grit-like. [PCG][*]

Absolute Humidity. *See Humidity, Absolute.*

Absolute Temperature. *See Kelvin Temperature.*

Absolute Value. The magnitude of a number without regard for sign. The absolute value of a number is never negative.

Accuracy. Consistency between a measured value and the true value.

Air Pressure. Force per unit area exerted by air molecules rebounding from a surface.

Aqueous Solution. A solution in which water is the solvent.

Area. The number of square units in a defined region of a surface.

ASTM D 1475. Standard Test Method for Density of Liquid Coatings, Inks, and Related Products.

ASTM E 691. Standard Practice for Conducting an Interlaboratory Study to Determine the Precision of a Test Method.

Atmospheric Pressure. The pressure exerted on the earth by the atmosphere. Standard atmospheric pressure at sea level is 1 atm = 14.70 psi = 101.3 kPa.

Barometer. A device for measuring atmospheric pressure.

Bell Shape. The symmetrical shape of a normal distribution curve. The center of the curve is the highest and is concave downward. Tails on either side approach zero as they move outward from the center and are concave upward.

Bias. A persistent positive or negative deviation of a test method average value from the assumed or accepted true value. [ASTM D 4790]

Bias Statement. In ASTM standards, the bias statement furnishes guidelines on the relationship between a set of typical test results produced by the test method under specific test conditions and a related set of accepted reference values. [ASTM E 177]

Binder. Nonvolatile portion of the liquid vehicle of a coating. When paint dries, the binder becomes part of the solid film. It binds the pigment particles together and cements the paint film to the material to which it is applied. [Painting/Coatings Dictionary] The amount of binder needed to completely wet a pigment is determined primarily by the pigment's particle size, shape, density, and chemical composition; the degree of polymerization; and the wetting properties of the binder. *See also Vehicle.* [PCG]*

Blast Cleaning. Cleaning and roughening a surface (particularly steel) with metallic or nonmetallic grit or metal shot (usually steel), which is projected against a surface by compressed air, centrifugal force, or water. [PCG]*

Blast Profile. *See Surface Profile.*

Caliper. A device for measuring length (or diameter) with a high degree of precision.

Canceling/Dividing Out. A shortcut when multiplying fractions whereby a numerator and a denominator of one of the fractions is divided by the same number.

Celsius Temperature. The temperature scale used in the metric system where water freezes at 0°C and boils at 100°C under standard atmospheric conditions; named after the 18th century Swedish astronomer who invented the scale; formerly called the centigrade scale.

Central Tendency. A value that the data tends to center around. Mean, median, and mode are measures of central tendency.

CFM. Cubic feet per minute.

Chloride Ion. Negatively charged ion (Cl^-) derived from the element chlorine. [PCG] *

Circle. A set of points in a plane all equidistant from a given point (the center).

Circumference. The distance around the outside (the perimeter) of a circle. $C = 2\pi r = \pi d$

Coating. (1) A liquid, liquefiable, or mastic composition that is converted to a solid protective, decorative, or functional adherent film after application as a thin layer [ASTM D 16]; (2) Generic term for paint, lacquer, enamel, etc. [Paint/Coatings Dictionary]

Coating System. A protective film consisting of one or more coats, applied in a predetermined order by prescribed methods. A coating system description in a specification may include surface preparation and quality control requirements. [PCG] *

Coefficient. The numeral that multiplies a variable. In the variable expression $3x^2$, the coefficient is 3.

Complex Fraction. A fraction where the numerator and/or the denominator are themselves fractions.

Confidence Interval. A range of values in which a parameter is likely to be found with a certain level of confidence.

Containment System. A system that includes the containment structure (i.e., containment walls, floor, supporting structure, and entryways); ventilation system (forced or natural air input ports, and natural or mechanical exhaust); and, in some cases, dust collection equipment. [Industrial Lead Paint Removal Handbook]

Conversion Factor. A fraction equal to one, with different units in the numerator and the denominator, that multiplies a quantity in order to change its dimensions (units).

Conversion of Units. The process of multiplying by one or more conversion factors to express a quantity with different dimensions (units).

Cylinder. A regular object with side walls perpendicular to a circular base.

Dehumidification. Removal of water vapor from the atmosphere inside a tank or other enclosed space by using desiccants or refrigeration methods.

Dew Point. The temperature at which air becomes saturated with water, that is, when the air is at 100 percent relative humidity. Below this temperature, moisture will condense and produce dew or fog. As air is cooled, the amount of water vapor it can hold decreases. Expressed another way, the point where the actual water vapor pressure becomes equal to the saturation water vapor pressure; any further cooling beyond this point normally results in the condensation of moisture. [PCG] *

DFT. *See Dry Film Thickness.*

Diameter. The distance across a circle, measured through the center.

Decimal Fraction. A fraction written as a decimal.

Decimal. A number consisting of a whole number followed by a decimal point (.) followed by the numerator of a fraction with a denominator that is a power of 10.

Denominator. The number below the line in a fraction; the number of equal parts into which the whole is divided.

Density. Usually the weight (or mass) per unit volume.

Dimensional Analysis. A process of analyzing the units in a formula to determine consistency.

Diluent. A volatile liquid which, while not a solvent for the nonvolatile constituents of a coating, may yet be used in conjunction with the true solvent, without causing precipitation. [PCG] *

Dispersion. The scattering of data about a central value.

Dry Film Thickness (DFT). Depth of cured film usually expressed in mils (0.001 inch) or micrometers (millionths of a meter). Standard methods for measuring dry film thickness are defined in SSPC-PA 2, ASTM D 1005, ASTM D 1186, and ASTM D 1400. [PCG] *

Dust Collector. A device or combination of devices for separating dust from the air handled by an exhaust ventilation system. [PCG] *

Epoxy Resin. Cross-linking resins based on the reactivity of the epoxide group. One common type is the resin made from epichlorhydrin and bisphenol A. Aliphatic polyols such as glycerol may be used instead of the aromatic bisphenol A or bisphenol F. [PCG] *

Equation. A mathematical statement of equality.

Error. The amount by which a measurement differs from the true value.

Error, Random. An error that is just as likely to result in a measurement being too large as too small.

Error, Systematic. An error that tends to push all measurements to one side of the true value. Systematic errors can often be corrected by adding or subtracting the same value from every measurement.

Estimating (Costs). The process of computing the approximate cost of a project.

Evaporation Rate. The rate at which a solvent or other volatile material is converted from a liquid to a gas under specified conditions. A solvent that has a low boiling point evaporates faster (is released from a wet coating faster) than one with a higher boiling point. [PCG] *

Exponent. A superscript written to the right of a number, letter, or algebraic expression that indicates how often the number, letter, or expression should be used as a factor. An exponent is a method of expressing repetitive multiplication. ($x^3 = x \cdot x \cdot x$). See Section 1.5 for more information on exponents.

Fahrenheit Temperature. The temperature scale where water freezes at 32° and boils at 212° under standard atmospheric conditions; named after the 18th century German physicist who devised the scale.

Film Thickness Gauge. A tool used to measure the wet or dry film thickness of a coating. [PCG] *

Fraction. A number written with a numerator over a denominator with a line separating them. The denominator tells into how many equal parts the whole shall be divided, and the numerator tells how many of these equal parts to take.

Geometry. The branch of mathematics dealing with points, lines, and shapes and the relationships among them.

Grain (Gr). A unit of weight historically equal to the weight of a grain of wheat. One grain is approximately equal to 65 milligrams.

Humditiy, Absolute. The weight of water vapor present in a unit volume of air; for example, grains per cubic foot, or grams per cubic meter. NOTE: The amount of water vapor is also reported in terms of weight per unit weight of dry air, for example, grains per pound of dry air. This value differs from values calculated on a volume basis and should not be referred to as absolute humidity. It is designated as humidity ratio, specific humidity, or moisture content. [ASTM E 41]

Humidity, Relative. The ratio of the actual pressure of existing water vapor to the maximum possible (saturation) pressure of water vapor in the atmosphere at the same temperature, expressed as a percentage. [ASTM E 41]

Improper Fraction. A fraction in which the numerator is not smaller than the denominator. Examples of improper fractions are 3/3 and 5/4.

Inorganic Coatings. Coatings based on silicates or phosphates and usually pigmented with metallic zinc. [PCG] *

Interest. Money paid to a depositor for use of the deposit; a charge for borrowed money, usually a percentage of the amount borrowed.

Intermediate Coat. One or more coats of paint between the primer coat and the finish coat. Sometimes called a mid-coat. [PCG] *

Iron Oxide(s). A red, brown, or yellow pigment, substantially rich in iron and marketed under a variety of names, such as red oxide, jeweler's rouge, Venetian red, ferric oxide, Indian red, red ochre, mineral rouge, Spanish oxide, and Turkey red. [PCG] *

Kelvin Temperature. An SI unit of temperature; a temperature scale that assigns the value of zero (0) to the coldest temperature possible. The temperature gradations are the same size as the Celsius gradations. The absolute temperature unit is the kelvin (K). Water freezes at 0°C or 273.15 K.

Kilopascal (kPa). A unit of pressure in the SI system of measurement. One kilopascal is approximately equal to 0.1450 psi.

Laminar Flow. Streamlined flow without turbulence.

Least Common Denominator. The smallest number into which all the denominators will divide evenly. The least common multiple of all the denominators.

Linear Equation. An equation where the exponent on the variable is one. The graph of a linear equation is a straight line.

Liter. A unit of volume in the SI system of measurement equal to 0.001 m^3. A liter is approximately equal to 1.06 quart.

Magnetic Pull-Off Gauge, Type 1. A dry film thickness gauge that uses a type of spring balance to pull a small permanent magnet from the surface of the painted steel. The magnetic force holding the magnet to the surface varies inversely as a non-linear function of the distance between magnet and steel, i.e., the thickness of the dry paint film. Type 1 refers to the gauge's designation in SSPC-PA 2. [PCG] *

Manufacturer's Technical Data Sheet. Sheet printed by the supplier of a product to provide instructions and information on its use.

Mass. The amount of matter in an object; a measure of a body's inertia (resistance to acceleration) or its gravitational attraction to another body. The fundamental mass unit in the SI system of measurement is the kilogram.

Material Safety Data Sheet (MSDS). A printed source of information about the hazards of materials, including coatings. Federal law, the Hazard Communication Standard, requires that an MSDS be published and supplied by the manufacturer of a hazardous material. The law also requires that employees have ready access to the MSDS in the workplace. The MSDS contains the following information: (1) product identification; (2) hazardous ingredients, their permissible exposure limits (PEL), and threshold limit values (TLV); (3) physical properties; (4) fire and explosion hazard data, such as flash point, lower and upper explosive limits, and fire fighting procedures; (5) health hazard data; (6) chemical reactivity/stability data; (7) spill and leak procedures; (8) special protection information; and (9) additional special precautions. [PCG]*

Mean. The sum of the measurements divided by the number of measurements; commonly referred to as the average value.

Median. The number from the middle of a list that is arranged in ascending (or descending) order. The median is not necessarily a number in the list.

Megapascal (MPa). A unit of pressure in the SI system of measurement. One megapascal is approximately equal to 145 psi.

Metric System. The international system (SI System) of measurement based on the fundamental units meter, kilogram, second, ampere, kelvin, mole, and candela.

Micaceous Iron Oxide. A hard, dense pigment of iron oxide which provides barrier protection by impeding the path of water and other corrosive agents in industrial coatings. It is used in protective coatings to increase resistance to blistering, delamination, and corrosion when formulated at a high pigment volume concentration. [PCG]*

Micrometer. A length equal to one-millionth of a meter (1 µm = 0.000 001 m); historically referred to as a micron.

Mililliter. A unit of volume equal to one-thousandth of a liter (1 mL =

0.001 L); a milliliter is also equal to a cubic centimeter ($1 \text{ mL} = 1 \text{ cm}^3 = 1 \text{ cc}$).

Mode. The measurement in a set of measurements that occurs more than any other measurement. A set of measurements may have more than one mode.

MSDS. *See Material Safety Data Sheet.*

Multiplication. Repetitive addition. Six multiplied by three means $6 + 6 + 6$.

Multiplicative Inverse. *See Reciprocal.*

Negative Numbers. Numbers that are less than zero.

Nonvolatile Vehicle. The coating binder dissolved or dispersed in the solvent (volatile vehicle) that remains to form a film after solvent evaporation.

Normal Distribution. A continuous bell-shaped probability distribution defined by equations not included in this book. In a normal distribution, the mean, the median, and the mode are the same number.

Nozzle Orifice Gauge. A tapered metal rod that can be inserted into the back of a blast cleaning nozzle to determine the size of its orifice and assess the extent of wear. The nozzle normally is replaced when the orifice size increases by 1/16 inch (1.6 mm). [PCG]*

Numerator. The number above the line in a fraction.

Order of Operations. An agreement among mathematicians as to the order in which operations (add, subtract, multiply, divide, etc.) are to be performed. Operations within grouping symbols (parentheses) are done first, then powers and roots, then multiplication or division in order from left to right, then addition or subtraction in order from left to right.

Organic Coating. Protective film with resinous (organic) binder. [PCG]*

Organic Compound. Any chemical compound that contains the element carbon in its molecular structure. [PCG]*

Organic Solvent. Liquid organic material including diluents and thinners that are used as dissolvers, viscosity reducers, or cleaning agents. [PCG]*

Outlier. A measurement that is outside the expected region; a measurement that is more that three standard deviations from the mean.

Paint. (1) Any pigmented liquid, liquefiable, or mastic composition designed for application to a substrate in a thin layer that is converted to an opaque solid film after application. Used for protection, decoration, identification, or to serve some other functional purposes. (2) Application of a coating material. [PCG]*

Paint Composition Formula. A recipe for making a particular paint.

Parallelogram. A quadrilateral with two sets of parallel sides.

Parts Per Million (PPM). Measure of proportional content (e.g., of small amounts of contamination). [PCG] 1 ppm = 1 μg/mL = 1 mg/L

Percent/Percentage. A fraction with a denominator of 100. The percent sign (%) means divided by 100. 100% = 1.

Perimeter. The distance around the outside of an object.

Pigment. Finely ground, natural or synthetic, inorganic or organic, insoluble particles that, when dispersed in a liquid vehicle to make paint, may provide color and other properties, including opacity, hardness, durability, and corrosion resistance. The term is used to include extenders as well as white or colored pigments. (Note: The distinction between powders that are pigments and those that are dyes is generally made on the basis of solubility, pigments being insoluble and dispersed in the material and dyes being soluble or in solution as used.)

Pigment Volume Concentration (PVC). Ratio of the volume of pigment to the volume of total nonvolatile material (i.e., pigment and binder) present in a coating. The figure usually is expressed as a percentage. [Coatings Encylopedic Dictionary]

Polyurethane. A coating binder material formed by the reaction of an isocyanate with hydroxyl-containing substances (polyols) to produce an organic polymer compound known as a urethane. Polyurethane products that are most suitable for coatings show good chemical, solvent, and abrasion resistance, and good gloss retention. (PCG) *

Polyurethane Coating. A coating based upon the reaction of an isocyanate-containing material with a polyhydroxylated co-reactant. Polyurethane coatings are available in single component (air-drying and moisture-curing) and in multipackage chemically curing or (catalyzed) formulations. (PCG) *

Power of a Number. Indicates the number of times a particular number is to be used as a factor. *See Exponent.*

PPM. *See Parts Per Million.*

Primer. First full coat of paint applied to a surface when a multicoat system is being used. Primers provide adhesion to a new substrate (wood, metal, masonry, or concrete), protect the substrate, and aid in the adhesion of additional coats of paint. The type and condition of the substrate and the painting system specified for a job affect the selection of the primer. Primers for steel work may contain special anti-corrosive pigments. [PCG] *

Precision. The degree to which several measurements of the same quantity agree with each other in terms of repeatability and reproducibility. The smaller the standard deviation of a set of measurements, the more precise the measurements are.

Probability. A measure of the likelihood that a particular event will occur. A probability of zero means that the event will not occur, while a probability of one means that the event will definitely occur. Probabilities fall between 0 and 1 and are often expressed as a percent.

Proportion. A statement of equality of two ratios.

PSI (Pound-force per square inch). A unit of pressure in the U. S. Customary system of measurement. One psi is approximately equal to 6.895 kilopascals.

Psychrometer. An instrument used to determine humidity and dew point. Sling psychrometers measure the wet and dry bulb temperature of air. With the aid of psychrometric tables, these measurements can be used to determine the dew point and relative humidity of the air. Electronic instruments are now available that provide results directly. [PCG] *

Psychrometric Tables. U.S. Weather Bureau Tables, originally published by the U.S. Department of Commerce, used to determine the relative humidity and dew point temperature from the dry and wet bulb readings obtained from the sling psychrometer. *See also Dew Point, Relative Humidity.* [PCG] *

Pull-Off Gauge. *See Magnetic Pull-Off Gauge, Type 1.*

PVC. *See Pigment Volume Concentration.*

Quadrilateral. A closed plane figure with four straight sides.

Quality Assurance. Steps taken to control measurement and testing processes through the implementation of policies, procedures, and corrective action to assess and assure the quality of analysis and results. [Industrial Lead Paint Removal Handbook]

Radius. A line from the center to the edge of a circle or sphere; a perpendicular line from the axis to the side of a cylinder; one-half of the diameter.

Range. The difference between the largest and the smallest numbers in a list.

Rate. A comparison of different quantities by division. A rate has dimensions while a ratio is dimensionless.

Ratio. A comparison of two quantities by division, usually expressed as a fraction, yielding a dimensionless number. *See Rate.*

Reciprocal. The multiplicative inverse of a number. A number times its reciprocal is equal to 1. For a fraction a/b, the reciprocal is b/a as long as neither a nor b is equal to zero.

Relative Humidity. *See Humidity, Relative.*

Root (of a Number). The inverse operation of raising a number to a power is finding a root. x is an nth root of y if $x^n = y$. Four is a square root of 16 because $4 \cdot 4 = 16$. Two is a cube root of 8 because $2 \cdot 2 \cdot 2 = 8$.

Equivalent statements are: $\sqrt[n]{y} = x$, $\sqrt{16} = 4$, and $\sqrt[3]{8} = 2$

Rounding. Approximating a number by using fewer significant figures.

Scientific Calculator. A hand-held calculator that has function keys such as log, 10^x, ln, e^x, x^y, sin, cos, and tan in addition to +, –, x, and ÷.

Scientific Notation. A method of writing a number as a number between 1 and 10 multiplied by the appropriate power of 10. The number 35,000 can be written in scientific notation as 3.5×10^4.

Shim. (1) A thin strip of non-magnetic plastic, metal, or other material of known uniform thickness used to calibrate coating dry film thickness gauges. (2) A strip of metal, wood, or other material employed to set base plates or structural members at the proper level for placement of grout, or to maintain the elongation in some types of post-tensioning anchorages. [American Concrete Institute Cement and Concrete Terminology]

Signed Numbers. A number greater than zero is positive and can be written with a + sign, e.g., +8. A number less than zero is negative and is written with a – sign, e.g., –6. If a number is written without a sign, it is assumed to be positive. Zero is neither positive nor negative.

Significant Figure. A digit with a known value.

Solvent. (1) Liquid, usually volatile, used in the manufacture of paint to dissolve or disperse the film-forming constituents. Since they evaporate during drying, solvents do not become part of the dried film. Solvents are used to control the consistency and character of the liquid paint material and to regulate its application properties [Painting/Coatings Dictionary]. (2) Liquid used to dissolve and remove oil, dirt, grease, soil, and waxes from metal surfaces. *See also Thinner, Solvent Cleaning*. [PCG] *

Solvent Cleaning. The use of organic solvents, detergents, alkaline cleaners, and steam cleaning to remove oil, grease, dirt, soil, and other, similar organic compounds from a surface. The specification SSPC-SP 1 *Solvent*

Cleaning is a consensus standard covering the procedures necessary for solvent cleaning of steel surfaces. Consult the document for specific details and requirements. [PCG]*

Soluble Salt Contaminants. Water-soluble inorganic compounds (such as chlorides and sulfates) that contaminate a surface or product. When soluble salts are present on a steel surface prepared for coating, they may cause premature coating failure, particularly by osmotic blistering. Soluble salt contaminants are sometimes referred to as "ionic contaminants" or "non-visible" contaminants. [PCG]*

Specific Gravity. The ratio of the density of a substance to the density of water. Specific gravity is a dimensionless quantity.

Specification. A word that is used in several ways in the coatings industry. The term "job specification" refers to the written, legal document, usually part of a contract, that precisely describes an item of work that is to be accomplished. Many technical organizations prepare documents called specifications that describe products, procedures, or conditions. For example, *SSPC Painting Manual Volume 2: Systems and Specifications* includes over a hundred frequently cited specifications covering surface preparation, abrasives, paints and paint systems, and paint application. Military, federal, state, and other agencies also have paint and related specifications. The Army and Navy also have documents called "guide specifications" (formerly called "type specifications"), which are called "engineering standards" by private industry. [PCG]*

Sphere. A three-dimensional object where every point on the surface is equidistant from a given point at the center.

Spot Measurement. The average of at least three gauge readings made within a 4 cm (1.5 inch) diameter circle. [SSPC-PA 2]

Spreading Rate. The area of surface covered per coat of paint at a specified dry film thickness per unit volume of coating material. Spreading rate generally is indicated by square feet covered per gallon or square meter covered per liter of paint.

SSPC-PA 1: Shop, Field, and Maintenance Painting of Steel. This specification covers procedures for painting steel surfaces after the selection of the coating material has been made. It does not cover surface preparation, pretreatments, or selection of primers and finish coats. SSPC-PA 1 is intended to be used for steel which, because of its exposure condition, will be subjected to corrosive attack, either from the weather or from the service environment, and where a high quality of cleaning and painting is essential. [PCG] *

SSPC-PA 2: *Measurement of Dry Paint Thickness with Magnetic Gages.* This specification describes the procedures to measure the thickness of a dry film of a nonmagnetic coating applied on a magnetic substrate using a magnetic gauge that is nondestructive to the film. [PCG] *

SSPC-Paint 25: *Zinc Oxide, Alkyd, Linseed Oil Primer For Use Over Hand-Cleaned Steel, Type I and Type II.* An SSPC composition specification for an inhibitive primer that has good wetting ability so it can be used on hand-cleaned steel.

SSPC-TR 3/NACE 6A192 (2000 Revision): *Dehumidification and Temperature Control During Surface Preparation, Application, and Curing for Coatings/Linings for Steel Tanks, Vessels, and Other Enclosed Areas.* A joint SSPC/NACE technical committee report on the use of dehumidification when painting enclosed areas.

Standard Deviation. A statistic that measures the dispersion of a set of numbers. See Chapter 6 for the formula for computing standard deviation.

Statistics. Values that are used to describe a set of numbers. The collection, description, and analysis of a set of numbers.

Surface Concentration. The weight of a contaminant (e.g., chloride) on each unit area of the surface; a measure of surface cleanliness usually given as micrograms of contaminant per square centimeter of surface area.

Surface Profile. The roughened surface that results from abrasive blast cleaning or power tool cleaning to bare metal. For steel, surface profile is a measurement of the peak-to-valley height of the roughness, often expressed as an average, and typically ranges from less than 1 mil up to 5 mils (25 to 127 microns). For wood and concrete, surface profile is simply the texture of the cleaned surface. [PCG] *

Swabbing. A method of extracting salt from a contaminated surface by washing the surface with cotton balls that have been dipped into deionized water. The salt contaminants dissolve in the water which is collected by the cotton swabs and then analyzed. Knowing the volume of water, the amount of salt collected, and the area swabbed, the concentration of salt on the surface can be determined.

Tech. Data Sheet. *See Manufacturer's Technical Data Sheet.*

Theorectical Coverage. Spreading rate of a coating to a completely flat surface, assuming no losses in handling.

Thinner. A volatile liquid used to improve the application properties of a coating, normally by reducing viscosity. A thinner may be a single solvent or a combination of solvent types. Often, specific thinners are required by the manufacturer of a coating to prevent damage to coating properties that may occur when an inappropriate thinner is used. *See also Solvent and Diluent.* [PCG]*

Thinning Rate. The amount of thinner allowed per unit volume of paint. This is usually expressed as pints per gallon, milliliters per liter, or as a percent.

Titanium Dioxide (TiO_2). A bright white, highly opaque, insoluble inorganic mineral used to impart hiding to coatings. [PCG]*

Topcoat. The last coating material applied in a coating system, specifically formulated for aesthetics and/or environmental resistance. Also referred to as a finish coat. [PCG]*

Transfer Efficiency. Percent of mass or volume of coating solids actually applied (transferred) from container to coated surface. [PCG]*

Trapezoid. A quadrilateral with one set of parallel sides.

Triangle. A closed plane figure with three straight sides.

Vehicle. The liquid portion of paint in which the pigment is dispersed. It is composed of binder, solvent, and anything dissolved in the liquid portion of the paint. *See also Binder.* [Painting/Coatings Dictionary]

Vehicle Solids. The amount of nonvolatile matter in the vehicle portion of paint, expressed as a percentage either of the total vehicle or the total paint vehicle solids. May be calculated by weight or by volume. *See Nonvolatile Vehicle.* [PCG] *

Vertex. The point where the two sides of an angle meet.

Viscosity. The quality or property of a fluid (e.g., paint) that causes it to resist flow. A high viscosity coating is thick; a low viscosity coating is thin.

VOC. *See Volatile Organic Compound.*

Volatile Organic Compound (VOC). (1) Any organic compound that reacts in the atmosphere with nitrogen oxides in the presence of heat and sunlight to form ozone; (2) Any organic compound (other than those designated by EPA as having negligible photochemical reactivity) that is emitted into the atmosphere during the application or curing of a coating. It is detected by reference methods such as EPA Method 24 or *ASTM D 2369 Standard Test Method for Volatile Content of Coatings.* [PCG] *

Volatile Vehicle. The solvent portion of a coating vehicle lost during film formation. *See Solvent.* [PCG] *

Volume. The number of cubic units in a defined region of space.

Volume Concentration (PPM). The weight of solute (salt) in a unit volume of solution. *See Parts Per Million.*

Volume Flow Rate. The volume of fluid (liquid or gas) flowing through a given cross-sectional area per unit time. Area (in square feet) times velocity (in feet per minute) of flow of the fluid equals flow rate (in cubic feet per minute).

Volume Percent Solids. The nonvolatile or solid components of a coating, expressed as percent by volume. This number can be used to calculate the amount of wet film thickness that needs to be applied to produce the desired dry film thickness of a coating. For instance, a coating with 50 percent volume solids will require twice as much wet film thickness as is desired in the dry film thickness, because half the volume of the material will volatilize. [PCG]*

Waterjetting. The use of water at high or ultrahigh pressure (pressures above 70 MPa/10,000 psi) to prepare a surface for recoating. High pressure waterjetting is cleaning performed at pressures from 70 to 210 MPa (10,000 to 30,000 psi). Ultrahigh pressure waterjetting begins at 210 MPa (30,000 psi). [SSPC-SP 12/NACE 5]

Wet Film Thickness (WFT). The thickness of a wet liquid coating film immediately after application and before evaporation of any volatile material from the film. [PCG]*

WFT. *See Wet Film Thickness.*

Zinc Dust. Finely divided zinc metal used as a pigment in zinc-rich and zinc dust/zinc oxide protective coatings for iron and steel. [Painting/ Coatings Dictionary]

Zinc-Rich Coating. Anti-corrosive primer for iron and steel. Zinc-rich coatings use zinc dust in a concentration sufficient to provide electrical conductivity in the dried film. This enables the zinc metal to corrode preferentially to the ferrous substrate, giving galvanic protection. [Painting/ Coatings Dictionary] Inorganic and organic zinc-rich coatings are described in SSPC Paint Specifications 20 and 29.

*Terms marked PCG are derived from SSPC's *Protective Coatings Glossary,* published in 2000.

www.ingramcontent.com/pod-product-compliance
Lightning Source LLC
LaVergne TN
LVHW022206200926
848481LV00006B/1861
9781575904757